This Way Up

MARK COOPER-JONES
JAY FOREMAN

WHEN MAPS GO WRONG
(and why it matters)

THIS WAY UP

MUDLARK

Mudlark
An imprint of HarperCollins*Publishers*
1 London Bridge Street
London SE1 9GF

www.harpercollins.co.uk

HarperCollins*Publishers*
Macken House, 39/40 Mayor Street Upper
Dublin 1, D01 C9W8, Ireland

First published by Mudlark 2025

5 7 9 10 8 6 4

Mark Cooper-Jones and Jay Foreman assert the moral right to be
identified as the authors of this work. If anyone else tells you they
wrote this book, don't believe them.

A catalogue record of this book is available from the British Library.

HB ISBN 978-0-00- 871027-9
TPB ISBN 978-0-00-871028-6
(Have you ever noticed every single ISBN ever begins with '978'?
Pick up any other book on your shelf right now and check. See?)

Printed and bound in the UK using 100 per cent renewable electricity
at CPI Group (UK) Ltd

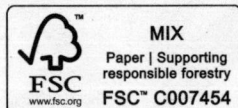

MIX
Paper | Supporting
responsible forestry
FSC™ C007454

This book is produced from FSC™ certified paper and other controlled sources to
ensure responsible forest management.
For more information on this, visit: www.harpercollins.co.uk/green
For more information in general, visit: en.wikipedia.org

Mark: This book is dedicated to my family.

Jay: No, this book is dedicated to *my* family.

CONTENTS

INTRODUCTION

In the words of famous American geographer Mark Monmonier, 'Maps are like milk, their information is perishable, and it's wise to check the date.' It can sometimes be difficult to find a date, but often, if you check the back . . .

Hang on. We should probably be clear about something before we go any further.

See, we're not actually historians. Or cartographers. Or cartographic historians. Or mapmakers or geographers or researchers or academics.

Our claim to cartographic authority is a video series on YouTube called 'Map Men', where the two of us (Mark Cooper-Jones on the left in a blue shirt, Jay Foreman on the right in a lighter blue shirt) sit behind a desk and talk about our favourite maps, occasionally dressing up in costumes and acting out the stories behind them in silly sketches. You don't need to have seen it; this book is not a sequel.

While it is true that our meaningful geographic credentials are limited, specifically . . .

Mark

- Got a degree in geography from the University of Durham (2:1)

- Was a geography teacher for three years (didn't like the students)
- Has a lapsed Fellowship of the Royal Geographical Society

Jay

- Spent his teenage years reading the *London A–Z*.

There's one factor that qualifies us to venture beyond our usual platform and politely elbow our way into the grown-up map section of a proper bookshop:

We *really* love maps.

And, if making silly videos about them for nearly ten years has taught us anything, it's that it turns out we're not alone.

Interest in maps appears to be on the rise. On a functional level, most of us are using maps more than ever before thanks to smartphones, while at the other end of the technology spectrum, the antique/collectable map market has been growing apace, with people increasingly drawn to their aesthetic, their stories and their sheer stare-at-ableness. And it's not just the wealthy art-collecting crowd who like to gawp. The subreddit r/MapPorn has 6.1 million members, putting it in the top 1 per cent of sub-reddits by size.*

There are lots of reasons why people enjoy looking at maps: beauty, curiosity, navigation, a desire to understand the spatial dimensions of a set of data, spotting your own house – but in our opinion, there's no map more thrilling than a map that's got some-thing wrong.

We should clarify what we mean by 'wrong'. Because if we're splitting hairs, every single map ever made is – by definition – wrong. To quote Mark Monmonier again, 'Not only is it easy to lie with maps, it is essential.'

---

* We recognise that there might be some wealthy art-collecting types on Reddit too, but surely not 6.1 million.

What he means is, no map can possibly be entirely accurate. A map's job is to take a bit of the real world and translate it into something useful, using their three defining features of scale, symbols and projection, each of which is a form of distortion. For a map to contain no distortion at all, it would have to be both three dimensional and at a scale of one to one, which is of very limited use.

What *we* mean by maps that have 'gone wrong' is maps with big, stinking, awful map blunders, like a country that's gone missing, or a fictional mountain range, or a mis-drawn border that crosses all sorts of boundaries – the sort of mistakes that could lead to the unfortunate map-user getting hopelessly lost. We love them because they provoke the question: *What on earth happened here?* And the answer is most often a fascinating story.

We also need to address the subtitle of our title's subtitle, the bit in parentheses: '*and Why It Matters*'.

On one level it should be fairly obvious. Wrong maps can have any number of grim consequences, depending on who is using them and why; from lorry drivers taking wrong turns into cul-de-sacs, to governments making bad decisions based on poor information, to mapmakers in authoritarian countries printing the border in a controversial position. And yet, in spite of their many inconveniences, we also think wrong maps have become something worth celebrating. Allow us, if you will, to introduce our light thesis to accompany all the messed-up map malarky.

[Steps onto dais, taps notes on lectern, clears throat]
Map use today is completely different from what it was only a very short time ago, because digital maps have, to all intents and purposes, stopped us getting lost.

The coming together of GPS technology, smartphones, digital mapping apps and speedy mobile networks has meant that, so long as we have battery life, we always know exactly where we are. Even the parking apps we're forced to use know our precise location at every moment.

On opening Google Maps,* we're immediately greeted with a flashing blue dot. 'No need to look up,' it assures us. 'You are precisely here. Just type your destination, and we'll guide you every step of the way.'

Where maps used to be an *aid* to navigation, something we could turn to to help us figure out where we are, now they *are* the navigation. There's no longer any human element, no orientation, looking around or cross-referencing landmarks. Digital maps have removed the need for any human cognition as they instruct us exactly where to turn, when.

At first glance: brilliant! Thinking is hard. And with so many other demands on our finite brain power (the state of our gut biome, the ending of *The Sopranos*, everyone else's infuriating cognitive biases), it's a relief to be able to outsource the chore of finding our way about.

At best, getting lost makes us cross; at worst, it makes us dead. So, what possible case could there be for reintroducing the risk of it happening?

Researchers in neuroscience have begun to answer that very question. It appears that our reliance on GPS, which for many of us stretches even to two-minute journeys to our local Post Office, has steadily dulled the hippocampus, the spatial part of our brain.

A University College London study published in 2000 looked at the brains of London's famous black cab drivers, whose profession famously requires an encyclopaedic knowledge of the city's streets learnt from years of full-time studying, equivalent to a law degree. The research showed that during their working life, taxi drivers' hippocampi became enlarged, giving them, on average, the largest examples of this part of the brain's anatomy of any profession, a fact they love to boast about while making small talk in traffic.

---

* Could be Apple Maps, could be OpenStreetMap, could be OS maps. Probably isn't.

Using our brains to find our way around is like flexing a muscle – a muscle which, if unused, begins to wither, a fact backed up by subsequent research that showed that the minute the taxi drivers hang up their gloves and retire, their hippocampi start to shrink back down again.

There's an unmistakable irony that by making fewer mistakes in where you're going, you decrease your understanding of your whereabouts. Being lost (assuming you eventually end up unlost again) does at least have the benefit of forcing you to think, solve the mystery of your whereabouts and improve your spatial perspective.

So, to return to our promised 'light thesis', it's this:

It's *good* to be lost.

Or at least, it's good to be forced to think about your surroundings and consider where we are in relation to the wider world.

And what better way to do so than to sample a sumptuous selection of fabulously faulty maps? Of course, reading this book won't get you lost (at least it shouldn't), but as more of us start to question the benefits of life under big tech, perhaps it might act as a timely ode to disorientation, a paean to the unknown.

A bit pompous, that. But then, Mark *is* incredibly pompous, and he wrote that bit.

In the following pages we've selected what we believe to be some of the very best wrong maps. Some of them are decades old, some are centuries old and some are so recent they're still being published today (or yesterday, if you're reading this tomorrow). They include colonial maps, collaborative maps, corporate maps, Soviet maps, pioneer maps, news maps and maps whose intended use was hijacked for a surrealist political movement – as well as a few others that weren't easy to write punchily into a list.

Some chapters tell the stories of unforeseen map errors, others of intentional errors. For one of them we even went on a little fieldtrip, which was hardly necessary, but old habits die hard, especially for ex-geography teachers.

For a pair of cartophiles like us, finding, researching and writing these stories while staring at the maps they're about has been endlessly satisfying. So if you're standing in a bookshop, flicking through the pages of this book deciding whether to buy it (and let's face it, if you've read this far the shopkeeper is probably now giving you evils), hopefully you'll find these stories a mixture of enlightening, fascinating and silly, while one or two might accidentally teeter into thought-provoking.

Above all, we hope the following chapters will lead you to agree with a view we feel more strongly about than ever before:

The worst maps are the best maps.

1

THE MAP THAT
DELETED A COUNTRY

'New Zealand is not a small country
but a large village.'
Peter Jackson

Our first map is called 'Björksta' ('byerkstaah', more or less) and was available at your local IKEA store for £30 in a 55 × 39-inch frame in a choice of black or silver, until it was hastily withdrawn and apologised for in February 2019.

For the benefit of those listening to the audiobook,* here's what IKEA's Björksta map *sounds* like.

> The ocean is white
> Sputterings of textured paint
> Form the shapes of land
>
> Creamy pastel shades
> With the occasional splat
> Of bright tangerine
>
> Don't look for meaning
> At splashes where colours meet
> It's random nonsense.

Or, by all means, you can google it.

While the Björksta map adds a nice dash of colour and worldly sophistication to the TV-dinner room, it also got the furniture giant that sold it into a heap of trouble, and upset 5.27 million perfectly nice people.

In February 2019 a Reddit user by the name of Jibbles666 spotted the map in their local branch of IKEA in Washington, DC. As they stared at the map, appreciating the sprinkles, they also became aware of a distant but unmistakable feeling of unease. Something was wrong. They began to sweat, their heart rate increasing. Why? The map seemed OK. Or was it? They couldn't quite tell. Maybe it was just the effect of all the meatballs – Jibbles absolutely loved meatballs and had eaten 55 as soon as they'd got in.†

---

* We've contractually agreed to record an audiobook. We currently have no idea how this is going to work re. all the maps. Will we even read this footnote out when we record it? Does one constantly stop, mid-sentence, to read out footnotes? Buy the audiobook now to find out what we did.

† This fact may or may not be entirely untrue and we apologise if the real Jibbles666 is vegetarian.

Then, with relief, they spotted it. Their eyes darted to the bottom right corner, towards an expanse of ocean, precisely where New Zealand should have been, but wasn't. Reverend* Jibbles took a photo of the map, and posted it to Reddit with the caption 'IKEA's map game is not on point.' This irresistibly sharable story quickly made it to proper news websites like the *New Zealand Herald* and BBC News, before eventually bursting out of the internet, reaching the 'And Finally . . .' sections of TV news broadcasts around the world. Within 24 hours IKEA was forced to withdraw the map from its stores and put out a sheepish statement, saying: 'IKEA is responsible for securing correct and compliant motifs on all our products. We can see that the process has failed regarding the product BJÖRKSTA world map – we regret this mistake and apologise.'

It was particularly bad timing for IKEA, because that very same month they were planning on launching their first store in New Zealand.

This is very funny.

It's perhaps unfair that IKEA got so badly picked on for their map, given that it was not the first to omit this particular 103,500 square miles of land. In fact, the webpage on which Jibbles chose to post the photo was a subreddit called 'MapsWithoutNZ', a forum with a vast and ever-growing collection of photos from across the world of world maps that don't believe in New Zealand. New Zealand's absence from world maps is a veritable phenomenon – a time-honoured tradition dating back thousands of years – and not just on rubbish maps where mistakes are expected, like hastily spray-painted murals in small cafés, children's toys, tattoos, or dull corporate logos with thick outlines. The country is unaccountably missing from otherwise perfectly detailed maps produced by a whole host of organisations who really should know better.

There's no New Zealand on the enormous world map on the wall of the Smithsonian Museum of Natural History, or the globe fountain at Florida's Universal Studios, or the cover of the 2016 textbook

* For all we know.

for AQA A-level geography, or either of the two board games *Pandemic* or *Risk*. Even the Flat Earth Society's logo has a dig – giving us two conspiracy theories for the price of one. We could go on.

We will. Almaty airport in Kazakhstan has a massive wall map in the customs hall with no New Zealand, which, by the way, resulted in a New Zealand visitor being detained for 24 hours when unable to point to the country of her passport in 2016. There was no New Zealand on the map produced for the Rugby World Cup in England in 2015, a time when New Zealand's All Blacks were the reigning world champions. And, our absolute favourite, there's no New Zealand on the massive United Nations logo on the wall of the organisation's New York HQ.

Most countries would be outraged at this sort of thing happening to them so often. Imagine China, the US, Australia, or – heaven forbid – Britain being left off a map even once; there'd be all-out war. But the Kiwis? Thankfully for IKEA, New Zealanders are masters of self-deprecation on a national scale. An afternoon spent scrolling through the MapsWithoutNZ comments section suggests that the overall mood of its citizens is at worst mild exasperation and at best a strange sense of pride at each time they've been forgotten. Like this is their *thing*.

Indeed, the New Zealand government are in on the joke themselves. Their website's error 404 page, which appears when a user tries to access a broken or non-existent link, features the caption 'Something is missing' with a MapWithoutNZ of their own. In 2018 New Zealand Tourism made a series of genuinely funny campaign videos about the 'great conspiracy theory to remove New Zealand from world maps', impressively featuring *The Lord of the Rings* director Peter Jackson (whose films were once lauded for putting New Zealand on the map – IKEA clearly didn't get the memo), actor Rhys Darby and the then prime minister Jacinda Ardern – although never in the same room, which is less impressive.

And so, all this begs the question, why *does* this keep happening? What is it about New Zealand that makes it so leave-off-of-maps-able?

A VERY FORGETTABLE COUNTRY?

Well, the name is a bad start. 'New Something' suggests a colonial outpost rather than a fully sovereign country that really should have been updated centuries ago. And 'Zealand' begins with Z, so you're already at the bottom of every list of New things.*

Indeed, there's evidence that there are enough geographically challenged people in the world who have trouble remembering that New Zealand is a nation. If you fire up Google and start typing the words '*Is New Zealand . . .*', the helpful suggestion for the most frequently typed rest of the sentence is '*part of Australia?*' It doesn't help that from far away the two countries' accents and flags are very similar and very similar respectively. Close your eyes and look at both flags. Can *you* tell the difference?

This has always been a point of contention for the Kiwis, destined to be forever unfairly and unnecessarily compared with their unfathomably large, continent-sized neighbour, which makes them appear from the other side of the world like a small pair of islands just off the Australian coast.

Helping even less is the fact that Australia already comes with its own medium-sized island dangling off the bottom – Tasmania, an Australian state comprising an island the size of Ireland. If New Zealand happens to be missing, amateur mapmakers might not notice, specifically because they spot Tasmania and go, 'That thing's New Zealand, isn't it?'

The country's forgettability explains why – should New Zealand happen to be missing from a map – the mistake is unlikely to be spotted and corrected. But why might New Zealand accidentally go missing in the first place? This one is a question of simple geography.

---

* It retains bragging rights over the New Zing Vaa Chinese takeaway on Ecclesall Road, Sheffield.

WHERE IS NEW ZEALAND?

That New Zealand could ever be considered small or near to Australia goes to show just how warped, unreliable and dependent upon context our perception of scale is. New Zealand is actually quite big. Bigger than Great Britain, for instance. It would take you roughly four months to walk the nearly 2,000 miles from Cape Reinga at its top to Bluff at its bottom (compared with less than two months from John o' Groats to Land's End). It's also nowhere near Australia. The shortest possible hop between the two countries is a four-hour flight crossing 1,400 miles, the same distance as London to Athens.*

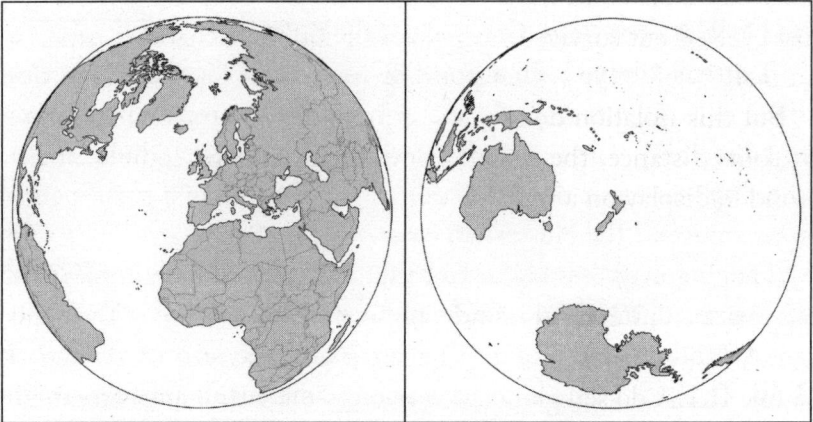

Left: As much land as possible, with Spain in the middle.
Right: As much ocean as possible, with New Zealand
in the middle.

New Zealand is not just far from Australia. It's far from *every-where*. A great way to demonstrate the extremity of this country's isolation is to divide the globe in half in such a way that one hemisphere contains as much land as possible, and the other as much ocean as possible. If you do this, you'll find New Zealand lurking

* For our Greek readers, that's the same as Athens to London.

dead centre of the ocean hemisphere – as if its position, as far as it can get, on average, from everybody else in the world, has been strategically chosen.

Indeed, New Zealand's isolation has played an enormous role in the nation's identity and history. New Zealand was one of the last places on earth to become infected with people, the Māori Polynesians only turning up at some stage between the 13th and 14th centuries, which is about 1978 in the grand scheme of things. This goes some way to explaining how the country remains so sparsely populated and the landscape so stunningly unspoilt, as the place famously contains more sheep than shoes.

It affects the wildlife too. Only on an island so uniquely detached from the rest of the world's land for millions of years could the kiwi – a rubbish bird with no wings or natural predators – have not just evolved but survived and risen to become the nation's national animal.

But this isolation does come at a cost. It makes New Zealand, by some distance, the most inconveniently located country in the world to display on a world map.

THE WORLD (MAP) IS FLAT

While IKEA do sell globes,* these are not practical for living room walls, diaries, classrooms, TV weather reports, shower curtains, board games, PowerPoint presentations or in-flight maps. For this reason, flat maps of the world, despite their necessarily fictitious nature, have outnumbered globes for centuries.

Performing a conversion of round planet to flat piece of paper requires three decisions. None of these have a 'right' answer, but the world has made up its mind about their favourite three answers, all of which are bad news for New Zealand.

---

* Aftonsparv, £7 at the time of writing. A soft plushy toy with a smiley face, featuring abstract colourful blobs for continents. Lacks contours.

1. Which way is up?

This might sound ridiculous. North being up is so ingrained it's hard to remember that this is an arbitrary convention followed by a mere vast majority of maps, with a minuscule handful of exceptions, usually produced to make a smug point, such as 'What if New Zealand were at the top for once?' North has been the standard for up on maps since European explorers first got their hands on a compass. This puts New Zealand down at the bottom of the map, as well as of your thoughts. Not a good start.

2. Which projection to use?

Any representation of a sphere converted into a flat plane necessarily has to be distorted in some way, either by being stretched, squashed or sliced. Try it yourself by drawing a smiley face on a balloon, stabbing it with a knife and spreading the remains out across the table. Doesn't look so happy now, does it? There are lots of available algorithms for projecting the curved surface of the earth onto a flat plane. At one extreme there's the Mercator Projection, which preserves accuracy of angles at the expense of scale, making Greenland balloon to the same size as Africa. At the other extreme there's the Gall–Peters Projection, which preserves accuracy of surface area at the expense of accuracy of shape, making Africa look twice as tall as it is wide.

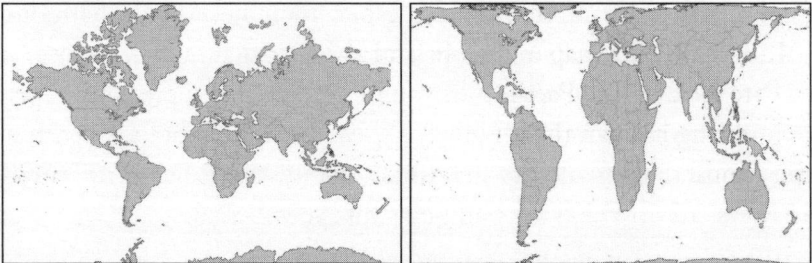

Left: Mercator Projection; Right: Gall–Peters Projection.
Accuracy-wise, they're as bad as each other.

Most world maps nowadays go for some sort of bendy compromise in the middle. But the Mercator Projection, despite its many tremendous flaws, including, most controversially, inflating northern Europe's size and sense of self-importance at the expense of Africa, is the projection of choice for that dependable, familiar, classic look. The Mercator Projection, despite looking nothing like the real Planet Earth, continues to this day to look *not* weird. This is further bad news for New Zealand, who, in this fictitious universe, finds itself dwarfed by nations in the north that are much smaller in size, including Norway, Sweden and Finland.

3. Where does the map stop?

Unless your flat map of the world is a repeating wallpaper pattern,* the map has to stop somewhere. This is a question of aesthetics rather than accuracy. A map that chops France in half, making it straddle both sides of the page, is a perfectly valid representation of Planet Earth, as long as every bit of France shows up at least once.† However, the standard, conventional, boring place to make the split is – and for many centuries has been – the middle of the planet's largest ocean, the Pacific, resulting in the layout we're all familiar with: Europe in the middle, Asia on the right, the Americas on the left. While this is partly, as with many things, arbitrary historical convention (Europe loves putting itself at the centre of everything), splitting the Pacific in half this way has numerous practical advantages.

1. It makes it a map of mostly land – the bit that humans want to look at. The Pacific takes up lots of valuable space, and slicing it down the middle is a great way to reduce its impact.

* IKEA do not sell this. They should.
† As a powerful member of the G7, they have repeatedly used their veto to prevent such maps from ever being published.

2. It shows countries' relative positions in terms of how most people will travel between them. The North Atlantic Ocean, for example, has 2,500 aircraft crossing over it every day, while long-haul flights avoid travelling directly over the Pacific if they possibly can. This makes the Pacific split an excellent map for showing most of the world's flight paths, shipping lanes and subsea internet cables.

3. The map isn't troubled by an international date line. The Prime Meridian at Greenwich, zero degrees longitude, is smack bang in the middle, making this traditional layout an excellent way of thinking about time zones. That the international date line is likely where it is *because* of this already popular layout is neither here nor there.

4. Perhaps most importantly, in the middle of the Pacific, it's possible to make a perfect vertical slice without injuring any major land masses. The same cannot be said of an Atlantic split, which various versions have to solve either by splitting Greenland in two, making the territory bulge outside the margins, or making Iceland appear twice – like Jay's grandpa managing to show up in his very wide black and white school photo twice by running round the back while the camera did its slow pan, which he got caned for but it was totally worth it.

It's worth noting that the Pacific split is not entirely universal. In some parts of the world, such as Japan, it's not uncommon to find Atlantic-split maps in classrooms, putting the Americas on the right, Europe on the left, and an enormously bulbous Pacific Ocean dominating the middle. Indeed, this is a very sensible way of understanding the world from a specifically Japanese point of view.

Or a New Zealand one, for that matter.

There's also a map that slices China in half purely for the benefit of the Americas (and Greenland), and it looks utterly ridiculous.

But these exceptions do a great job of proving the rule. In these Asian and Oceanian countries, Atlantic-split world maps are seen

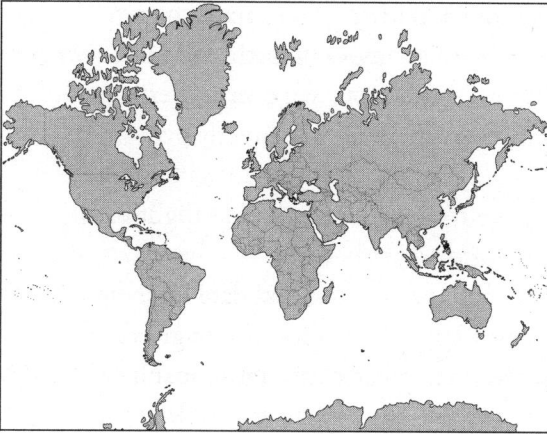

The very familiar, very boring but very practical, Pacific split.

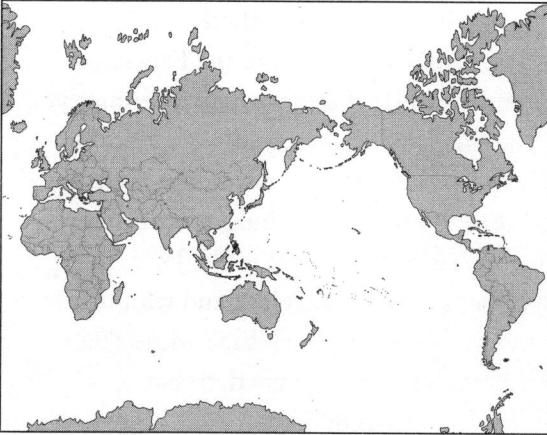

The Atlantic split, often seen in Japan and China.

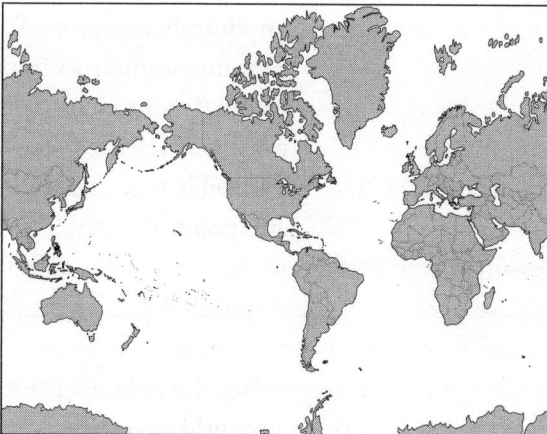

The Asia split. Looks dreadful.

alongside the ubiquitous Euro-centric maps, not instead of them. A map with Portugal on the left looks jarring and unfamiliar to pupils in Portugal in a way that a map with New Zealand on the right doesn't to pupils in New Zealand.

And so, for all of the above reasons (or reasons to the left, depending where the page split lands in this book), we have our default, standard, near-universal flat map of the world. North is up; the north is stretched; the Pacific is hidden round the back. And this is how New Zealand ends up, most commonly, in its precarious and dangerously forgettable position: extra low down, extra small and extra close to the edge.

Shoved down here in this bottom corner, the Kiwis find themselves vulnerably easy and tempting to crop out of the map altogether, either by accident, which is disrespectful, or convenience, which is disrespectfuller. This is what happens in a lot of the cases posted on the MapsWithoutNZ Reddit page. Many of them just miss the country out because it's simply too hard to fit it in without having to make everything else smaller.

While careless cropping is one possible explanation, if not an excuse, for the majority of MapsWithoutNZ, it doesn't explain what went wrong with IKEA's offending Björksta map. New Zealand wasn't beyond the edge of the frame; there was a massive expanse of empty ocean in the bottom right where New Zealand could and should easily have gone. So, what else could have gone wrong?

MAKE IT SIMPLE, STUPID

It's completely reasonable for mapmakers to lose some fine detail for the sake of clarity, especially if they're using thick outlines, limited colours, low-resolution printing or plasticine.

For this reason, some world maps have no New Zealand because it naturally disappears when you zoom out far enough – especially on a simplified, artistic or impressionistic map that exists for fun rather than navigation. One example from the Reddit page is a world map on an oven tray made out of chicken nuggets, with the perfectly good excuse that they don't make nuggets small enough for New Zealand (and if they did, they'd be burnt to an inedible crunchy crisp).

But this doesn't explain IKEA's mistake either. The Björksta map has no thick outlines, and the rest of the map contains plenty of detail, showing many islands that are much smaller, crinklier and fiddlier than New Zealand, such as Sicily and Jamaica.

One can only assume that a deliberate decision to remove the massive pair of islands that make up New Zealand was taken by a human person. So how exactly can it have happened? While IKEA acknowledged that the mistake was bad – and, they claimed, accidental – and that they'll never do it again, they were very light on the details as to which individual was responsible for the mistake and how they made it. So, it's time for us to don our deerstalkers, grasp our magnifying glasses and try to piece it together.

It is a cartographic inevitability that if New Zealand is missing from a map, plenty of other places will be missing too. And sure enough, a closer inspection of the Björksta map reveals that New Zealand is far from the only sin. Start by zooming into any area you're familiar with – in our case, Britain – and you'll notice that the shapes are, in fact, *all* wrong. Shetland is missing, and Anglesey has been engulfed by a blob that looks nothing like Wales.

The rest of Europe is no better. Scandinavia – where this map apparently comes from – is a wobbly blob, with Denmark smushed into Sweden, and all of Norway's famously lovely crinkly edges have been sandblasted and made unnervingly smooth. The more you look, the worse it gets. Over in Asia, the Black and Caspian seas are both missing, making Turkey disappear, and across the Atlantic, not a single one of the North American Great Lakes is wet. All this suggests that every inch of coastline on the Björksta map must have been traced, probably in a bit of a rush, manually by hand.

A close-up look at a familiar corner of the world reveals that
the Björksta map is geographically rubbish.

This gives us our first clue as to how New Zealand might have
got away. But to truly understand how it happened, we need to get
inside the map designer's head. Let's imagine our Swedish graphic
designer sitting at their laptop in their sunlit conservatory. Using
our knowledge of mapmaking software, a little speculation and
the process of elimination, we've boiled it down to four possible
scenarios as to what they might have been thinking.

Scenario 1

*Tum te tum te tum, time to open up Adobe Illustrator and import a
vector file with the basic shapes of the world map for me to transform
into beautiful art . . . Whoa! That's a lot of layers! Coastlines, landmass,
colour fills, city symbols, text labels. I don't need any of that, I just want
the basic outline of the world map for me to trace. Delete layer . . . delete
layer . . . delete layer . . . Oh, what? Why is each landmass its own
separate layer? This is all over the place! And it's making my laptop run
so slowly! Delete, delete, delete, delete, delete, Whoops! Too far! Just got
rid of loads of countries by mistake. I'd better put them all back. Undo,*

Undo, Undo. Errr . . . Is that all of them? Yeah, looks pretty much like a complete world map to me. Great! Done.

Scenario 2

Tum te tum . . . Right, I've got Microsoft Paint open on my laptop, and an atlas from my dad's house on my knees. Time to draw the whole thing by hand using the trackpad. Tum te tum te tum, squiggle squiggle. Oh no! New Zealand. My least favourite country. Still never forgiven them for pipping the Swedish team to the gold medal in the K-4 1,000-metre canoeing at the 1984 Olympics. I'm going to deliberately leave a gap on this world map where New Zealand should be. That'll show 'em!

Scenario 3

Tum te tum, making a map, making a map. Oh, that's the doorbell. I wonder who that could be. Hello? Aaaaargh!!! A big purple monster! I'd better run back inside the house! Pant pant pant. Oh no! He's following me back into the house? Why did I leave the door open? I'd better jump into the nearest cupboard. Pant pant pant. Phew. Safe at last. Oh no! He's opened the cupboard and started making a loud gurgling noise. Get back, purple monster, or I'll throw my hot coffee on you! . . . Splosh! . . . Wow! Didn't expect that to work so effectively or instantly. I'd better clean up all that melted purple monster off my wooden floor. Right. Where was I? Oh yeah. Time to delete New Zealand off this world map.

Or, perhaps the most temptingly believable scenario might just be . . .

Scenario 4

Tum te tum te tum. Oh look, my phone's ringing. I'd better answer it. Hello? Oh, hello IKEA client. Yes, it's nearly finished. Yes, I'm aware of the popular internet meme of New Zealand missing from world maps. No, I didn't know you were launching your first store in New Zealand

*within the next few months! Hahahahahaha!!! Yes, I'm definitely up
for that. That would be really funny! I'm sure someone will spot it
pretty quickly. And what a brilliant bit of PR that would be – a quirky
bit of harmless fun at the expense of a nation who've proven time and
again that they're perfectly capable of taking a joke. Yes, I imagine that
would lead to people mentioning IKEA on news articles, websites,
Reddit threads and books about maps for many years to come. Of course
I can! Won't take a second. Let me just click here, drag down to here,
aaaaaand . . . deleted! Any more countries you want me to get rid of?
Nope? OK, no worries. Bye!*

IT'S NOT JUST NEW ZEALAND

Perhaps the main thing we learnt from our long and careful stare
at the Björksta map is that New Zealand is not special. While
the comments in the MapsWithoutNZ forum suggest that they
are uniquely unfortunate in being forgotten by the rest of the
world, in this regard, New Zealand is, unusually for New Zealand,
not alone.

Even within MapsWithoutNZ, the overwhelming majority of
the maps are missing more than NZ. For just a few examples, a
geography textbook has no Ireland, the boardgame *Risk* has no Sri
Lanka, the Flat Earth Society logo has no Britain, and the UN
logo on the wall of the UN has no medium-sized islands at all
(albeit due to the thick slabs of metal it's made from, which probably gets it off the hook.)

There are *many* countries around the world that regularly fall
victim to map simplifications and omissions, each with their own
hallowed traditions of complaining online when they go missing,
including, but far from limited to, Madagascar, Iceland, the
Philippines, Taiwan, Malta, Cyprus, and almost every Caribbean
and Pacific island. The list gets longer when you include dependencies and *parts* of countries – the Channel Islands, the Isle of
Man, Sicily, Sardinia, the Balearic Islands, Michigan's Upper

Peninsula, Corsica and many *many* more. It gets longer still when you include bits of water that also go missing. Lough Neagh, the Great Lakes of North America and the Black Sea all get regularly filled in in the name of simplification – or ignorance. And don't forget the entire continent of Antarctica – which, of course, *most* world maps do!

So, perhaps the question should not be 'Why does New Zealand go missing all the time?', but rather 'Why do we hear so much about New Zealand going missing all the time?' What is it that makes the maps in MapsWithoutNZ so much more easily shareable than the ones in MapsWithoutNauru, MapswithoutMauritius or MapswithoutNovayaZemlya?

What it all comes down to is that of all the places that have the misfortune of regularly going missing, New Zealand is the funniest. Partly because it's an entire nation state, partly because its size and isolation make for a comparatively easy game of spot the difference, partly because they speak English, but mostly because New Zealand has taken to it with such good humour and made their omission their own.

As it transpires, going missing from maps all the time has really worked out in New Zealand's favour. IKEA's lamentable Björksta map – while embarrassing for the company – ironically did more to bring New Zealand to the world's attention than an accurate, responsibly produced map with a correctly imported vector file ever could.

Forgetting New Zealand is, it turns out, a brilliant way to remember New Zealand.

2

THE MAP THAT GUESSED

'The Mississippi River will always have its own way.'
Mark Twain

*'Just a short back and sides please, I'm not feeling adventurous.
And could you trim my moustache?'*
Also Mark Twain, that same afternoon

DRAWING A MAP OF A VERY BIG PLACE

Imagine your job is to make maps charting the 'New World' of North America in 1750.

If your imagination is any good, it will probably go something like this:

*** *Cue a rising xylophone scale. The world goes wavy like a dream sequence on* Sesame Street.***

Your name is John Mitchell, a British mapmaker living in Virginia, one of the 13 British colonies of British North America, and you're panicking. You know you're very good at making maps – one of the best – but there's a lot of land on this new continent, and, truth be told, you've barely seen any of it and can hardly have been expected to. Every day there are new borders, frontiers and features to be marked up for the information-hungry generals and politicians, and there are plenty of mapmakers piling in on this cartography gold rush. Some of the maps are OK, some of them are cheap rip-offs. You've started working on a map of your own,

and you'd very much like for it to be very good, so you've decided
you'll try very hard for it to be so. But the scale of the challenge
feels, at times, overwhelming.

'It's too BIG!'

You wake with a start, imagining vast expanses of land, another
one of your night terrors. You find your thumb and soothe yourself
back to sleep.

You dream of a crow's-eye view of the continent you inhabit.*
Questions start flooding in as your limbs fill with anxiety. Who
owns what? What is where? What will become of this land? The
eastern section is currently scarred by violent wars between three
nations, all of which are technically located three and a half thousand miles across the Atlantic Ocean.

First, there are the 13 British-owned eastern colonies. They're
your people and you love the king. But there are also the Spanish
somewhere off to the south (Florida) and west (the Mississippi).
And as if that wasn't bad enough, the French are here too. Everywhere around you, in fact, and you perceive them as the very real
threat that they very much are.

At this point in your dream, you decide to draw a rough sketch
mapping out where these three rivals are based in case you ever
come to write a book featuring your map and it would be helpful
to give readers a visual sense. Because you attempt this doodle in
your dream, it's below average.

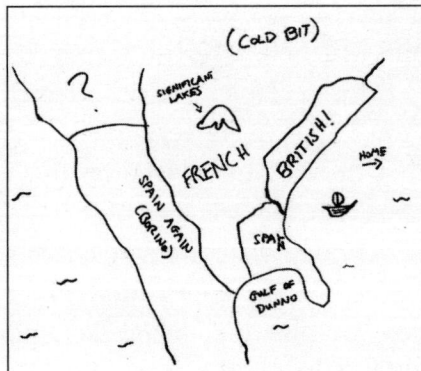

* NB: You're now in a dream within your dream.

By area, most of this continent lies out to the west, but the way you experience that space within the crow's eye map-mare is frustratingly blurred and at the edge of your vision. Perhaps this is because even your lucid mind can barely imagine the scale and extent of the vast plains stretching all the way to California (which you believe to be an island). Cartographically speaking, the west feels like a problem for another day. A day, say, after your own death.

Pleasantly, this stress-dream soon morphs into something altogether less cogent involving a russet potato, your dead grandmother and the word 'Thursday'.

When you wake,* you double-down on the job at hand, funnelling fear and anxiety into furious cartography until, in 1755, you finally complete your map. It looks like this:

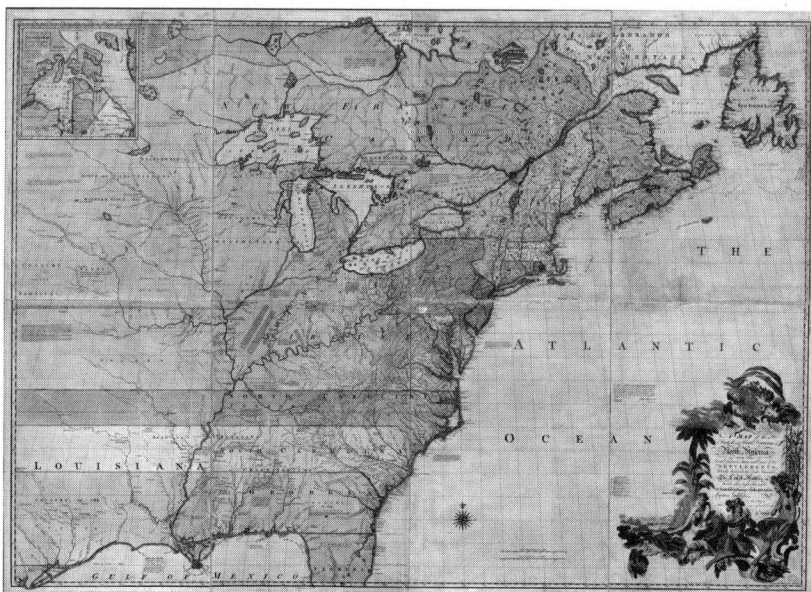

---

* You're now just in the first dream again. As in, yours, about what it would be like to be John.

At the bottom right is a splendidly long title garnished with cherubs and botany, indicating that this is A Very Important Map. Between the Bay of St Lawrence, the Great Lakes and the Mississippi River is a huge amount of detail, showing lands belonging to or claimed by different colonial powers. The map also manages to show lots of detail around the Hudson Bay, which had recently been surveyed and was worth throwing in. But because it was a good few miles north-west of the edge of the map, you've employed a method used by many cartographers at the time (and to this day), and put it in a neat little inset – a box with a frame around it at the top left. (This decision will have consequences in a few paragraphs' time.)

As you stare lovingly at the finished product, the dread and panic begin to subside, and a new emotion takes hold: pride. Partly from a well-earned sense of achievement, but also pride at something yet to happen, something to do with destiny and fulfilling your patriotic duty. You just know deep down that *your* map will be one of The Greats, used for decades by statesmen and dignitaries alike, a pivotal tool in the shaping of the New World, destined to play a key role in global history.

You believe that role will be wholly positive.

****The map begins to appear wavy, and the* Sesame Street *music returns. You (not John Mitchell, you you) come to. It's today again, your imagination is no longer under scrutiny, and the tense this story is being written in is about to revert jarringly to the past.****

John Mitchell's map was – upon release and for too many years afterwards – *the* map of America. Nobody thought of it as perfect; the town of Worcester was accidentally named Leicester, for example. (Or, as one or two of the settlers were beginning to call them, War-sester and Ly-sester.) But it did exactly the job it was meant to: help the British understand their territory, keep their territory and limit incursions by the French. Having been commissioned by the Crown, it was a colonialist's map, designed to support the British cause in whatever ways it could. Unlike maps of the North America we recognise today, Mitchell's map didn't even get close to the western coast because – Europeanly speaking – everyone lived in the east at the time. The west, vague as John had left it, hardly mattered a jot. The map spawned numerous editions, and when John Mitchell died in 1768, he was nothing but happy about his big, lovely, clever, super-famous map.

A NEW NATION

In 1775, twenty years after Mitchell's map was published, a nasty war broke out. What had started as a small argument over a tea party had escalated into a full-scale bid for independence from Britain by a small, feisty start-up country known as the United States of America.

By this time, who owned what in North America already looked radically different from the time of the first publication of Mitchell's map in 1755. Britain, for instance, had found itself in territorial credit following the Seven Years' War with, among others, France, and was now administering a good chunk of what was later to become Canada. No doubt Mitchell would have been delighted to learn that both French and Spanish power on the continent were on the wane, mostly due to the difficulty of fighting wars both in the New World and back home in Europe. He would, however, have been less delighted to find out that Britain, rather than taking advantage of this retreat, was slowly *losing* a war against an upstart republic.

By the 1780s, conversations were being had between the British and a crew of Founding Fathers about where these so-called 'United States' should specifically be and not be.

It was time for the Mitchell map to help make history.

THE MITCHELL MAP MAKES HISTORY
(AND REVEALS ITS FATAL FLAW)

The American Revolutionary War (or Illegal Terrorist Rebellion, depending whose side you were on) culminated in 1783 at the Treaty of Paris, where the very most important people the British and Americans could find all gathered round a big candlelit table to thrash out the details of who owned what. The New World was about to get a new international border between the British to the north, and the USA to the south.

Of course, to draw an international border properly, you need a big, detailed map. And thankfully, the British had just the thing. An enormous version of the Mitchell map was now spread across the candlelit table. Sure, it was 30 years old, but a jolly good map's a jolly good map; if anything, its age gave it a trusty gravitas.

You'd imagine the treaty bit at the end of any war to feel frosty and bitter; the losing side grunting monosyllabically, with the winners smirking at one another, perhaps even breaking out into chants of 'U-S-A, U-S-A' as it was being signed. The truth at the Treaty of Paris was that Britain knew its time was up, and although it was keen to ensure its land north of the United States was as big as possible, the strongest displays of emotion were reserved for when one of the American negotiators asked a waiter to cook his Chateaubriand well done. For the most part proceedings were amicable; plucky British negotiator David Hartley, for instance, was said to have been close friends with opposite number Benjamin Franklin. The Treaty of Versailles this wasn't.

Armed with Mitchell's prize map, these history-making officials got their red pens out and drew their red lines all across it in a bid to find a decent agreement.

As previously mentioned, John had always had a keen eye for delineating borders between what had, in 1755, been British and French territory (with a view to favouring British claims). As a result, the border between Quebec and northern American states such as Maine, New York and Vermont was well marked, with minimal room for squabble.*†

Where Mitchell had been altogether more hazy was the northern American border to the west of Quebec, a substantial portion of which was now under scrutiny by negotiators negotiating the exact placement of the Americo-British border. But, of course, they had no choice but to do the best they could with what lay in front of them, even if it meant making one or two assumptions about what may or may not actually be there.

By the time the Treaty of Paris had been signed and ratified in September 1783, the line they had agreed upon went as follows:

Ah, hello. Gosh, how awkward to be a line like this. Um, look, I need to just squidge my way across the west, if you don't mind – have you all agreed the best route? Go on then, I'm all ears. Start in the Bay of Fundy? That's not a real bay, is it? Oh right. Of course it is, HAHAHA. Then straight up the St Croix, *as the actor said to the bishop!* HAHA. Sorry, no, you're right, deadly serious. Double back down those hilly bits – you know me, love a good hike! Then zip across the 45th parallel – *Quebec on my right, Vermont on my left* – I can remember that. Through the Great Lakes, very sensible, just what I'd have done, splish splash splosh, etc, etc. Then up Rainy River to the northwest point of Lake of the Woods then directly west to the Mississippi – Em-*eye*-ess-ess-*eye*-ess-ess-*eye*-pee-pee-*eye* as my geography teacher used to scream at me,

* Oddly, the British territorial shoe was now on the other territorial foot. Where in 1755 Quebec had been French; and America British, now in 1783 Quebec was British and America was about to be American, meaning that what had once been territorial biases favouring Britain on Mitchell's map were now precisely the opposite.

† Franklin did, in fact, ask outright if the USA could include *all* of British North America, including the Canada bit, just for simplicity's sake. The British considered this request, then declined.

HAHAHA! Good, so we hit the Mississippi and from there the rest is Spanish anyway, as they say!

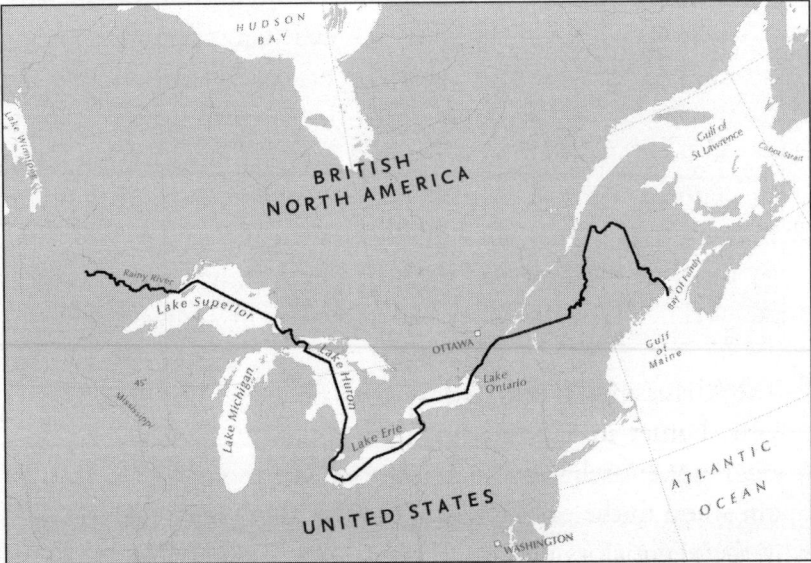

Here is the region the line had reached shown on a zoomed-in section of the Mitchell map. For reference: we've circled Lake of the Woods and added a massive arrow pointing to the Mississippi.

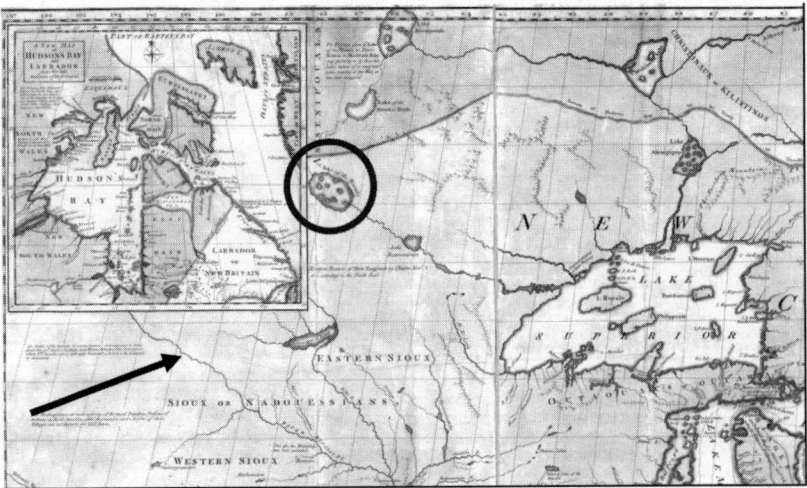

And, for context, here is that section in relation to the rest of North America.

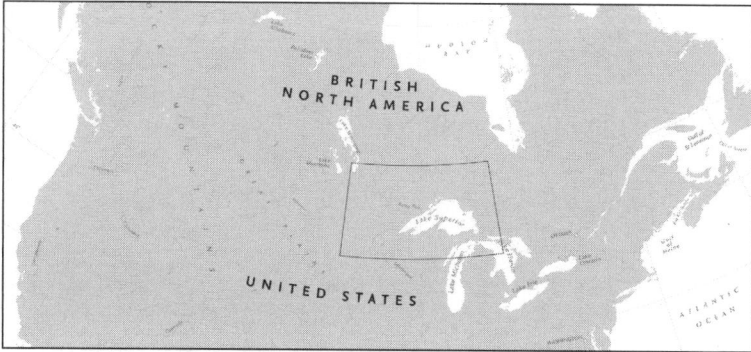

The Mississippi, being both a river and huge, was *the* key geographic border in 18th-century North America.* It was broadly (though not entirely) agreed that everything west of it belonged to Spain – and to the east, everything belonged to not Spain. This is why the negotiators needed to reach the Mississippi to mark the end of the British–US border in 1783. But unfortunately, on Mitchell's map, there was a bit of a problem – the location of the Mississippi at the required latitude was hidden under the aforementioned box in the top left.

The negotiators, presuming that the river must have continued *somewhere* under Mitchell's box, came up with an agreement that would hopefully avoid any future confusion. The new international border line, having travelled north-east up the Rainy River, was to continue, in the words of the treaty itself:

> to Lake of the Woods; thence through the said lake to the most northwestern point thereof, and from thence on a due west course to the river Mississippi.†

---

* Perhaps with the exception of the Atlantic coast.
† By the way, if 'river Mississippi' sounds odd, you can blame the British influence. Americans always write 'river' *after* its given name, whereas Brits write it *before* (River Thames, River Avon, River Piddle vs Ohio River, Sacramento River and, of course, Mississippi River). Nobody knows why.

In short, the western end of the British–American border was to come in a completely unknown location under the Big Box.

Had Mitchell actually visited this area, or spoken to someone who had, he'd have discovered two things:

1. Lake of the Woods was not a small round blob as depicted, but significantly larger and more complex.
2. The Mississippi was, perhaps for the only time in its history, not long enough.

WHAT FRANKLIN KNEW

'Psst, come here.'

Benjamin Franklin beckoned his pal, John Adams, away from the British crew to the dark corner of the room where he was giggling and nursing a large whisky.

'What? Why?' scowled Adams. 'No! The cheeseboard is over here, and I don't want the gouty Brits to eat it all.'

'Come *here!*' Franklin insisted. His smirk wider than a Great Lake, reluctantly forcing Adams out of his chair as curiosity overcame curd.

'They're using the Mitchell Map,' whispered Franklin feverishly.

'Yeah? So?'

Franklin tapped his nose, and swigged some whisky slightly too fast, pouring a bit onto his arm and brushing it off frantically with his other.

'I know something they don't know,' he half-sang to the tune of a long-established playground taunt.

'What?' urged Adams, intrigued.

'Well, you know they keep talking about this line west of Lake of the Woods connecting with the Mississippi?'

'Yeah.'

'And they think the source of the Mississippi is under that Big Box somewhere.'

'Yeah.'

'It's not.'

'Shut the front door. Where is it?'

'Dunno. Somewhere south.'

'How do you know?!'

'Oh, you know. Been studying the work of Jonathan Carver from the 1760s. He wandered around there, drew some maps. Most people haven't seen them but . . .'

Franklin revealed the corner of some folded paper in his top jacket pocket.

'Oop!' he boasted.

'But . . .' started Adams, the cogs beginning to turn. 'This could help us . . .'

'Oh yes,' said Franklin. 'The British must never see this. They love that our western border is so clearly hemmed in by the Mississippi. But! If the Mississippi doesn't actually reach our border with the Brits as they suppose it does . . .'

'It could open the door to westward expansion.'

'Indeed. But also, if the British get their hands on this, they might push for a border further south at the Mississippi, and we wouldn't want that, would we?'

'No, Benjy, that would mean *less* land for us. Which is *bad*.'

'You're a sharp biscuit,* John.'

Adams now begged to see the map, but with the lightest flick of his fingers Franklin dismissed him back to his cheese chair.

'All in good time, John. All in good time. And remember, mum's the word.'

John nodded, and as he turned to glance at Mitchell's map once more, he couldn't help but let out a scoff at its inadequacy.

---

* The word 'cookie' only came into use 25 years later, in 1808.

A SOLUTION IS NEEDED,
A SOLUTION IS FOUND

Despite the Americans knowing full well that they were signing up to a mathematically impossible border west of Lake of the Woods, the truth was that at the time of the Paris agreement in 1783 this part of the continent didn't overly matter. No American settlers lived near Lake of the Woods, and Minnesota – the US state that today sits south of border – wouldn't exist for another 75 years.

In the early 1800s, however, things started to change. The Louisiana Purchase in 1803 ceded 828,000 square miles of land around the Mississippi from France to the USA for just $15 million, the same amount that Keanu Reeves was paid for *Matrix Reloaded* (before add-ons). This meant that America had now successfully expanded westwards, and the border between Britain and the US had just got a lot longer. As American settlers began to spread further and further west in search of land and prosperity, they needed to be sure which country they were set- tling in. The Treaty of Paris now badly needed an update.

Helpfully, a man called Henry Schoolcraft found the source of the Mississippi in 1822. Turns out it lay directly south of Lake of the Woods at Lake Itasca. So that cleared that up – no Mississippi west of Lake of the Woods. Whoops!

But the other thing that needed resolving was the precise loca- tion of the north-west corner of Lake of the Woods, which had also been mentioned in the original treaty, despite no one knowing at the time what the lake looked like. Interestingly, this was the more challenging job, because it turned out Lake of the Woods actually looked like this:

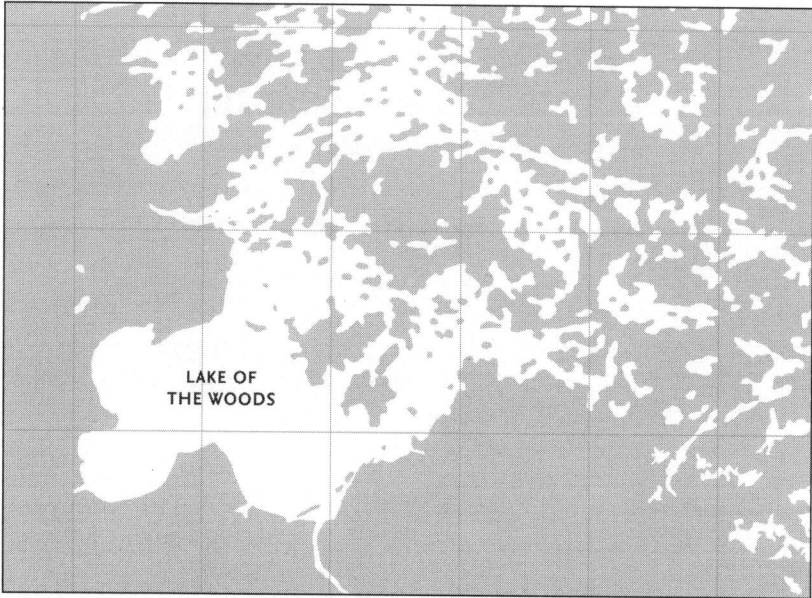

Exactly what counts as 'the north-west of' this splattering
of wobbly shapes?

And so, that same year, a team was sent to locate the north-west
point of the lake. They failed. Another team, sent in 1824, came
up with four possible places but couldn't agree which it was.
Finally, in 1825, a German astronomer drew some mathematical
lines across a map, putting everyone out of their misery and revealing that the painfully sought-after spot was the tip of a narrow
peninsula called 'Angle Inlet'.

Only now, the problem was that this north-west point turned
out to be a lot further north than everyone had assumed it would
be. The Americans didn't mind, but the British very much did – a
border further north meant less land for *them*. As a result, both
sides agreed the border from the Angle Inlet should drop directly
due south down to the 49th parallel, the line of latitude that circles
the earth exactly 49 degrees north of the equator. This was more in
keeping with where they'd assumed the north-west point would be
in the first place. From there, the border could continue west along
this line if it wanted to.

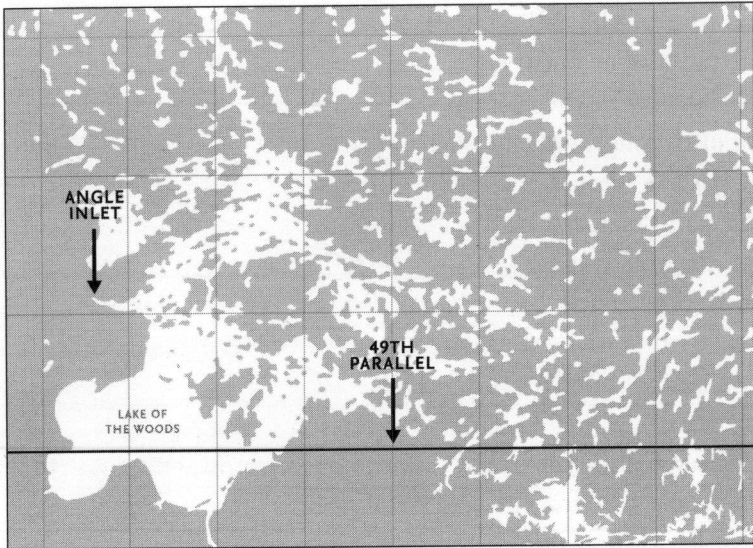

After all that careful line-wibbling across the north-eastern border of British North America and American North America, the border from the northwestern point of Lake of the Woods would now simply go as follows:

Down. Then left.

THE 49TH PARALLEL

Throughout the 1800s the internal borders of North America continued to change at a blistering pace. Which mostly meant, the United States kept getting bigger.

As more and more American settlers spread west, and more land and states were bought and acquired, the map delineating the States' northern border continually needed updating. In all the negotiating at the Jay Treaty of 1794, the London Convention of 1818 and the Oregon Treaty of 1846, the border's route never got any more creative than 'just follow the 49th parallel'.

Both sides ultimately felt it was easier and more agreeable to simply stick to the imaginary mathematical line used to resolve the issue of the Mississippi River's shortcomings. The result is that the border between Canada and the USA today is a map of two halves. On the eastern side it's twisted and intricate, a reflection of its long-contested European history (as mapped by Mitchell). To the west, however, it famously careers straight along the 49th parallel for over a thousand miles; a reflection of the fact that this area was far less populated by European settlers and was not fraught with contentious historic territorial fights (except those with Native Americans, whose own territorial claims were, as we know, roundly ignored).*

However, just because the western bit is straight, this doesn't mean it's sensible. Far from it. In fact, it's one of the most ludicrous bits of border in the world.

First, just the physical act of staking it out in a pre-satellite age was no small chore, as teams of surveyors were dispatched to lay out 900 markers across this 1,260-mile stretch of the 49th parallel. Tasked with keeping the line dead straight, they undoubtedly tried their very best, but nobody's best is good enough when you're schlepping across hundreds of miles of rivers, lakes, mountains and forests with a compass and a few bags of twine. Owing to

---

* If there's one thing that does unify the varied US–Canada border today, it's the hugely eccentric 'no touching zone'. Like a pair of seven-year-olds savagely protecting their own desk space from one another, Canada and the USA have each agreed to keep a ten-foot gap of nothingness either side of their border to clearly mark it across thousands of miles. And that's not just a rule applying to man-made structures; it includes trees. All trees, and small bushes too. No substantial flora is allowed in a 20-foot space along the entire 5,500 miles of border, including the Yukon–Alaskan bit. If something woody starts growing, there's a patrol, they're out there and they're hacking it down before it . . . you know, grows, and . . . er, confuses the . . . beavers, about where the border is. It's one of those jobs where you think, this could actually be a waste of not just money, but also time, nature, space, thought, effort and every other conceivable form of energy that goes into this. Either way, it's a thing, making the US–Canada border the longest deforested line in the world. At least it's clearly marked.

understandable human error, the border today is a series of mostly tiny wibbles in and around the 49th parallel, often straying as much as 800 metres from its intended line of latitude.

These wibbles are, however, never practical or sensible wibbles; only random. The line was always utterly dedicated to its founding principle never to deviate around anything that stood in its path, even if it would make nothing but total sense to do so. River meanders were cut off rather than followed, settlements were bisected and entire communities were cut off from their homeland.

In fact, they still are.

STRANDED NORTH AMERICANS

The borderline bungles of yesteryear are not just an amusing quirk of history, they continue to have a profound impact for people who live on or near the border today.

The Northwest Angle, otherwise known as Angle Township or simply 'The Angle', is the northernmost part of the main joined-up

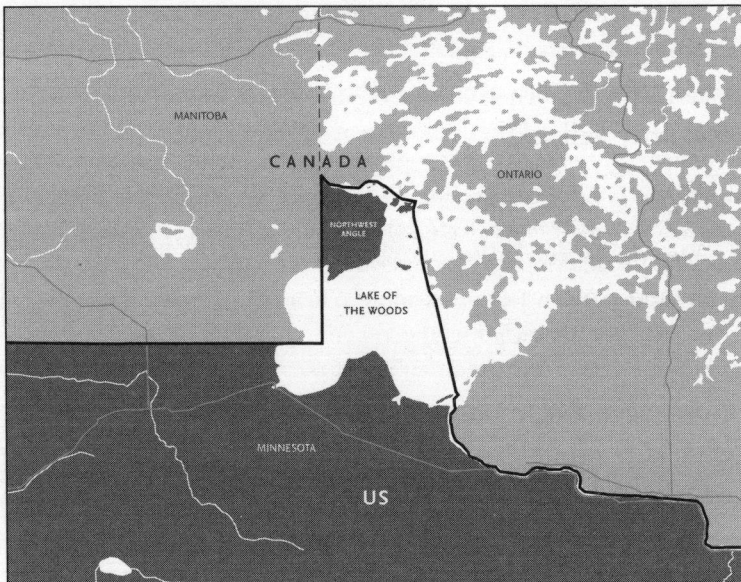

bit of the USA – i.e. not including Alaska – and it's a region that fell victim to the Big Box problem from Mitchell's map.

When Britain and America agreed that the border should – in keeping with the spirit of the Paris agreements – drop straight *down* from the north-west point of Lake of the Woods to the 49th parallel, The Angle became an immediate casualty of common sense.

By all laws of geography, it should be in Canada, as its only land borders are with the Canadian state of Manitoba. But it's not, it's in American Minnesota. Because why bend the line you're drawing to correct one problem, when you can use a ruler to create another? By drawing a straight line from the north-west corner of Lake of the Woods directly south to the 49th parallel, they cut off The Angle from the rest of the USA except by means of boat. Oh, and boat isn't an option unless you bring your own, because no vehicle or commercial ferry services operate here.

According to the 2010 census, the area is home to 119 people (99.2 per cent of whom were white – you do the math). It's an area of 600 square miles, most of which is water. For work, the majority fish; for recreation, it has a five-hole golf course.

To reach The Angle, your best option is to drive over 60 miles from Minnesota, through Manitoba, back to this isolated bit of Minnesota. Then you go to a border crossing 'reporting booth' known as 'Jim's Corner' and press a button to speak to a customs official. They will ask:

- Who are you?
- Are you a serial killer on the run?
- Should America cede this bit of land to Canada and make everyone's lives a lot easier?

And, if you pass, you can enter/exit.

Most of the time this formality is merely a minor hassle. After all, these are two super-friendly North American nations who get along (most of the time), and Angle residents tolerate the administrative inconvenience. But when international borders were closed for months on end during the Covid-19 pandemic, this cartographic quirk became a genuine problem. Key supplies began to dwindle, and the fishermen had nowhere to sell their fish and make a livelihood.

And so, in winter 2020, the town came together to create by far The Angle's best form of connectivity yet – a seasonal ice road across the frozen lake, which they now carve out every year to give this quirky geographic anomaly an additional transport lifeline.

Of course, they could have been saved the hassle if the men with the maps had made the obvious and sensible decision to simply draw the line *around* the region rather than insisting it plummet directly south to the 49th parallel. But they didn't, and the upshot is a community so cut off from their motherland that they've been forced to drive across miles of ice just to survive.

Perfectly normal.

The Angle is the original silly cut-off casualty of the US–Canada border's storied history,* but it is not the only one.

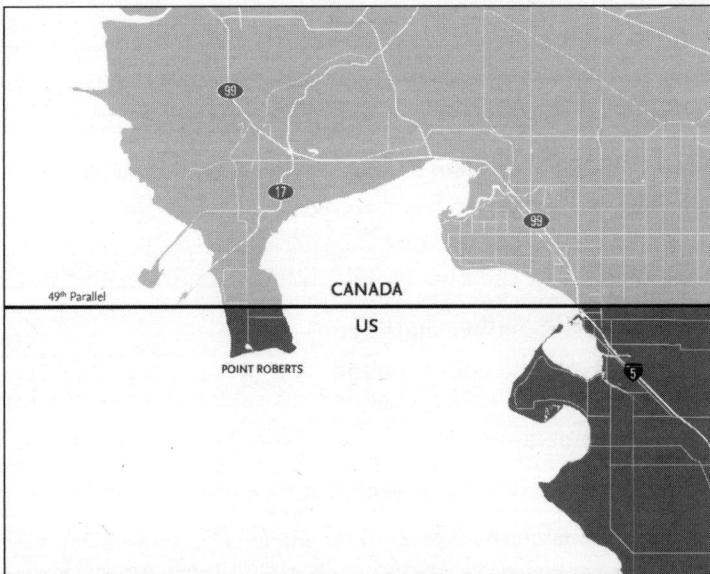

* Or, as we could have referred to it throughout this chapter, 'the Canada border'. It only has one.

Point Roberts sits just south of Vancouver on the very western edge of the continent. It juts out at the end of a peninsula into the Strait of Georgia, and should by all laws of common sense belong to Canada.

Unfortunately for Point Roberts, the 49th parallel runs through it – the border line that always had to be straight no matter what it encountered. Once again, even though it would have been incredibly easy at any point in the last two hundred years to bend the border round the peninsula and give the land to Canada, nobody ever has. As a result, if the good children of Point Roberts have to cross an international border four times a day to get to a school in their own country, then that's the price you pay for appealing to geometry rather than geography.*

MITCHELL'S LEGACY

In many ways, this chapter has been unkind to John Mitchell. His map was, by the standards of the day and what was known at the time, an incredible achievement, and it feels a tad unfair that it gets its own chapter in a book about maps that went wrong.

But for two wealthy, supposedly orderly nations, the US–Canada border is a basket case of baffling border bungles, and some of that is undoubtedly due to the unfortunate shortcomings of Mitchell's map. Had he simply placed his Big Box in a different bit of the map – say, the bottom left, the border between these two nations might look very different.

For one, the cutting off of Angle Township was a direct result of the fact that Lake of the Woods (whose shape he didn't know) had to be used as a co-ordinate to mark the western end of the

---

* The Point Roberts quirk is made all the crueller by the fact that the only place the 49th parallel does deviate is just to the west of here, to dog-leg around Vancouver Island, which the British somehow managed to secure for themselves.

border in 1783, which wouldn't have been its fate had his Big Box not been where it was. But it's possible that the straightness of the central western stretch of the border is also a consequence of Mitchell's map, *despite* that line being agreed long after the Treaty of Paris. Had the two sides not subsequently decided to bring the border from Lake of the Woods down to such an artificial point as the 49th line of latitude, would they have decided to continue the border in such an artificial way thereafter? Perhaps. Perhaps not.

Ultimately, the greatest cruelty to Mitchell is that his map was used in the 1783 Paris negotiations at all. After all, this was a man who had dedicated his life to proudly supporting the British Empire in North America, and who had created his map to strengthen British territorial claims against the French. That it was this same work that was then used to facilitate the loss of the 13 colonies to a new, breakaway, upstart *republic*, no less, is an irony he should be grateful he never lived to see.

Though he would, no doubt, have been delighted to have inconvenienced the process.

3

THE MAP OF THE WRONG CITY

AUTHORS' PAIRING NOTES

If you're partial to some light background music while reading, then this chapter comes with the authors' own selected pairings: why not try some Serge Gainsbourg, Edith Piaf, Erik Satie or perhaps even Miles Davis to set the tone?

If you have Spotify, snap the handy QR code below for a carefully curated playlist.

And, if you don't enjoy listening to music while reading, we'd highly recommend you stick on John Cage's *4'33*. That, too, is a perfectly suitable accompaniment for the subject.

'We will not lead; we will only detonate.'
Guy Debord

10 A.M., NOVEMBER, PHONE CALL

Jay: What's the plan today?

Mark: We're meeting for lunch at 2 p.m. I sent you a map of where we're going.

Jay: Yeah, I saw that. You sent a map of Paris by mistake.

Mark: Nope. I sent a map of Paris on purpose. We're having lunch in the Champs-Élysées.

Jay: In Paris?

Mark: No, London, obviously.

Jay: I'm so confused.

Mark: It's a map challenge. We'll both head into central London, but we have to find our way to the restaurant using this map of Paris. If we end up in the same place, we meet for lunch.

Jay: That makes no sense.

Mark: I know! Fun, isn't it?

Jay: Why are we doing this?

PLAN DE ~~PARIS~~ LONDON

Mark: Research.

Jay: Will you tell me what this is all about when we find each other?

Mark: *If* we find each other.

12.21 P.M., NOVEMBER, LONDON

The following passages are taken from transcripts of Jay and Mark's real audio diaries of this experiment.

Jay: I really have no idea what's going to happen today. I'm heading towards central London on the Tube, trying to decide where I ought to get off. I think my best bet is to head to the international station at St Pancras, since that seems to be the only place on this map of Paris that has a London equivalent.

Mark: OK, so I've just got off a train at London Bridge station. Knowing as I do that London Bridge is south of the river, I'm going to say that I am at Gare d'Austerlitz right now.

(*Map data from Open Street Map.*)

Which doesn't sound very French. Anyway, thankfully, the route from Austerlitz to the Champs-Élysées looks relatively straightforward, so I'm going to follow the directions on this map as well as I can on the streets of London, and hopefully meet Jay for a lovely sandwich at 2 p.m.

And if you're wondering what on earth we're doing and why, then you can blame Guy Debord, a dead radical French geographer who absolutely loved to get lost.

A RADICAL GUY

In 1957 a Marxist French philosopher and filmmaker called Guy* Debord had lots of thoughts and theories in his head, and felt he was ready to turn these thoughts and theories into a movement with a name, and perhaps even some followers. He had, in fact, previously contributed some of his thoughts and theories to another movement, which someone else had started, called the Lettrists (the movement, not the someone else). But it hadn't really caught on, and everyone he mentioned it to said 'Lettrists' was a rubbish name and he should start up on his own.

Guy was inclined to agree. After all, he had all the necessary credentials for founding a revolutionary philosophical school of thought: he had proved himself a profound thinker, theorist, activist, academic, revolutionary, vagrant and drunk. He even wrote articles for publications with names like *The Belgian Surrealist Journal*, which showed just how dedicated to abstract thought he really was. And if there was one thing that really got Guy's intellectual juices flowing, it was capitalism. In particular, undermining it; exposing its vice-like grip on all aspects of society, culture, art, work, play and politics.

For Guy, the problems of capitalism extended far beyond the baseline Marxist critique of economic inequality and into the very

* Pronounced 'ghee'.

fabric of people's lives. Everywhere he looked he saw a rampant capitalist monoculture that had poisoned the lake of human existence, suffocating creativity, originality and true subjective experience. Society was full of 'imbeciles', unwittingly trapped by the routines capitalism had built for them, blinding them to the possibility of anything different.

These routines included actions you or we might consider a fairly fundamental part of existence:

- Going to work*
- Buying a sandwich
- Buying some clothes in River Island†
- Consuming art, films, books, ideas, most of which were produced for commercial interest and therefore had no intrinsic artistic value
- Navigating a city using all the typical routes designed to funnel you between various capitalist experiences.

Guy had also come up with a rather neat term that summed up the way in which capitalism worked to keep us in check and obscure the possibility of a different lived experience. He called it 'The Spectacle', a mass-media-, advertising- and consumer-led artificial reality that had pulled the wool over society's eyes and, in short, ruined everything.

Equally valid reactions to this include 'That's a bit extreme' and 'Guess he's got a point'.

In July 1957 Guy found himself wondering if he was the man to un-ruin everything via the medium of founding a movement.

And one thing was for sure. This movement would contain some novel thoughts on how to use a map.

* Guy believed wage labour was a euphemism for slave labour. Guy never had a day job.
† Founded in 1948, River Island did exist at the time. There's no record of Guy ever visiting one.

12.40 P.M., LONDON

Mark: Slow progress so far. I exited London Bridge station by my nearest exit and immediately took a right as my map told me to. I then took another right which has unfortunately sent me straight back into London Bridge station by a different entrance. I will re-evaluate when I'm on the other side of the station.

I've also just realised that I'm perilously close to our publisher's offices, where I expect our editor might be working right now. I've decided not to see if he's about or to tell him what I'm doing, in case he has second thoughts about our arrangement.

Jay: It's been 20 minutes and I'm still in St Pancras station. I was struggling to decide which exit to use, as none are labelled on my map of Paris. I blame the scale. Anyway, I decided to ask a station attendant if he knew the way to the Champs-Élysées and he told me to go to the Eurostar check-in just around the corner. He seemingly didn't understand that I'm meeting Mark for lunch *in London*.

Anyway, I've decided not to pay hundreds of pounds to go to the wrong city, and instead have been forced to do something I've never ever done before in London, which is to ask for help at a tourist information office.

Mark: Embarrassing admission. I walked back through London Bridge station only to realise I'd actually read the map wrong, and I wasn't meant to take a second right after the first right. What this means is I need to go back through the station again, for a *third* time, and go back to the road I was on and turn right until I hit a bridge. Which, knowing the area as I do, isn't going to happen. How do I know if I'm doing this right?

Jay: I just had a genuinely unbelievable interaction with a staff member from the tourist office. I showed him my map of Paris, pointed to the Champs-Élysées, and asked him which was the best exit from this station to walk there. He then proceeded to give me actual, confident directions, which, of course, I must now follow.

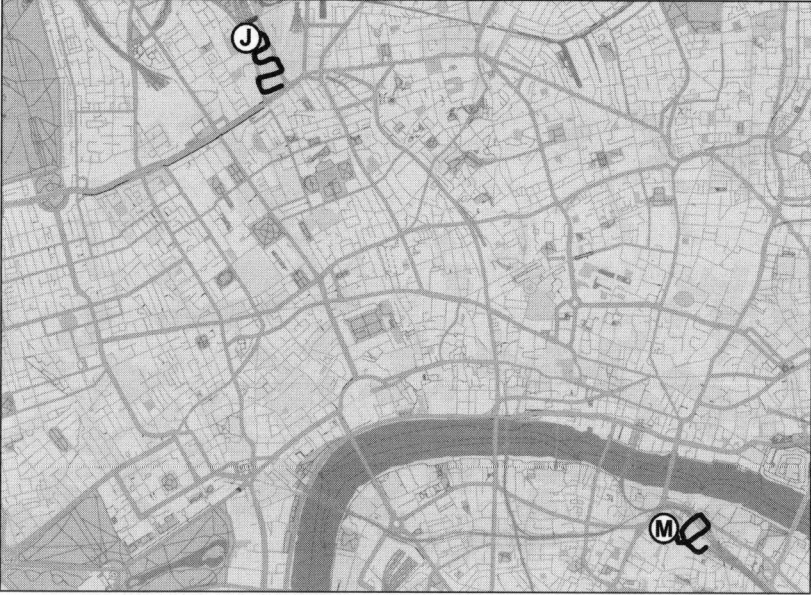

(Map data from Open Street Map.)

THE ECCENTRIC EIGHT:
ADVENTURES IN ITALY – PART I

One fine afternoon, just a week after Bastille Day, Guy Debord and his seven intellectually subversive friends were travelling by train to the small town of Cosio di Arroscia in northern Italy. It was just a hop, skip and a jump over the French border, and excitement was in the air.

The Eccentric Eight – as we shall call them for now – were off on an adventure. Not a normal adventure, mind, like the sort that involves treasure, smugglers or camping.*†‡§ No, this was a very different sort of adventure. It was an adventure of the conscious mind into the enticing realm of social theory, structures of oppression and the true meaning of art.

Guy Debord had a round face and small glasses, and his slightly puffy cheeks made him appear deceptively unthreatening. Michèle thought Guy's cheeks were cute, and had once made the mistake of saying so. Michèle was the strong-headed, quiet girl in the group with short, cropped hair that made her look a bit like a boy. She had complimented Guy's cheeks on their wedding day, which had almost ruined the whole thing, because it turned out that Guy was so deeply self-conscious about them that if anyone expressed the slightest affection in their direction, he'd accuse them of rank insincerity, which is exactly what happened. Thankfully, they got through it and were now a happy and formidably intellectual couple.

Guy was staring out of the window at the picturesque grain fields rolling past, his gaze drawn by a tractor ploughing its predictably boustrophedon pattern. He recognised that to most this was an idyllic image of the bucolic French countryside, but try as he might to enjoy it, Guy couldn't shake a disdain for the imbecile slavishly driving up and down, ad infinitum, doubtless motivated by nothing more than a fistful of pathetically meaningless French francs.

'If I were driving the tractor,' mused Guy, 'I'd at least drive it around the field in random directions so as to experience ploughing in a new and surprising way.'

He wrote that down in a scrapbook, and just as he was getting

* Editor: What's going on?
† Authors: We're just trying to make it really obvious that this section is a parody of Enid Blyton's famous children's adventure books, *The Famous Five* and *The Secret Seven*.
‡ Editor: Oh, I see. Yeah, that's really brilliant, actually. Definitely keep going with it.
§ Authors: We will.

excited by the idea of reversing a combine harvester through the Place de la Concorde, his attention was diverted by his close friend the artist Asger Jorn. Asger had pulled a set of oil paints out of his bag, and was suddenly painting a canvas in the middle of the train carriage. For a moment, Guy worried he was painting a traditional landscape of the countryside, which would have been grounds for expulsion from the movement before it had even got going. But when he moseyed over to see the work for himself, he could see that Asger was graffitiing an insipid Parisian watercolour he'd bought from a flea market with his own fantastically garish slogans, to update and vastly improve the work.*

'Brilliant', thought Guy. 'Asger can be my right-hand man.'

Guy looked around the carriage. Besides him, Michèle and Asger, there were Walter Olmo, Elena Verrone, Piero Simondo, Giuseppe Pinot-Gallizio and the English artist Ralph Rumney. None of them had bought train tickets.

Guy was feeling optimistic. Soon, they'd be across the border in Italy and their radical adventures could finally begin.

12.56 P.M., LONDON

Jay: My strategy so far has been to continue to ask people for help, and, would you believe it, I happened to pass a small group of people who were speaking French to one another. In my best French I asked them how to find the Champs-Élysées, but I think they thought I was playing some sort of prank on them. It was a little bit awkward. The woman laughed at me and the man stared blankly, saying what I think translates to 'I do not understand the premise.'

I'm pretty sure asking tourists for directions in my own city, in their language, is what the spirit of this challenge is supposed to be all about.

---

* Subverting art in this way was to become a feature of Guy's movement.

Mark: It might sound like an obvious thing to say, but it's surprisingly difficult to use a map of Paris to find your way around London, because it turns out the streets are unhelpfully different.

I'd been growing restless about the presence of the river on my right when it should, according to my map, have been on my left, but then I saw St Paul's Cathedral across the river. Consulting my map again, I could see that a large cathedral was indeed supposed to be across the water from me, so I concluded I must be holding the map upside down, and was travelling in precisely the wrong direction. I've turned the map the right way up – with south now at the top – and committed to a second 180-degree turn in just 20 minutes. I *must* now be heading the right way.

I'm now walking back towards London Bridge, however, and everything feels familiar rather than 'new' and 'surprising'.

Jay: The map suddenly seems to be on point, because it suggested I need to pass two metro stations before I take a left. And sure enough, I've gone past Euston Square station

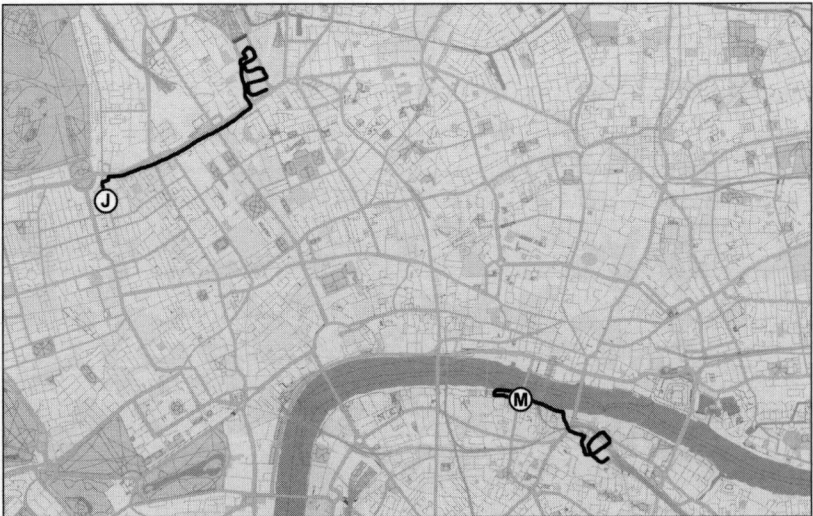

(*Map data from Open Street Map.*)

and then Warren Street station. Just like the map then said I would, I've come to a large round building with streets coming off at all angles. The sign says 'Great Portland Street station', but everything about my map tells me this must be Place de l'Opéra.

Things are looking up . . . I *think*.

THE ECCENTRIC EIGHT: ADVENTURES IN ITALY – PART II

'But what are we going to call ourselves?' pleaded Piero, who was excited to know what Guy had come up with.

The Eccentric Eight were sitting around a long table; Guy was at the head, Asger Jorn opposite, while the other six – and 13 empty wine bottles – filled the space between them.

Guy very deliberately folded a piece of Italian ham around a solitary gherkin, the last of his breakfast. He hoped the pause made him appear confident and in control, though truthfully he wasn't 100 per cent sure people would like the name he had in mind. He brought the ham-gherkin up to his nose, and inhaled its scent dramatically, before he finally announced:

'The Situationists'

and popped it in his mouth.

Silence, as the group took this in. Guy felt sure it was a much better name than the Lettrists, but the immediate reaction was unclear. After a moment, a few people began nodding in a mixture of thought and what Guy decided must be agreement. Piero, however, was frowning.

'"The Situationists"?'

'Yes,' Guy replied. 'The Situationists International.'

'Cool,' said Ralph. 'I like it.'

'Thanks,' beamed Guy, relieved.

'But what does it actually mean, "Situationists International?"' continued Piero.

This, Guy was prepared for.

'It is our primary duty in life to create "situations".'

'Ahhhh,' whispered Elena, to herself.

Guy went on: 'We must replace apathy and passivity with experience, encounters and spontaneous acts of play; resist everything that the city's traditional design, architecture and authorities intend for us to do.

'We will expose the city's bourgeois structure – like how our wide streets have been designed purely for loathsome cars, and our buildings have been built for businesses and rich people, excluding the working classes. We will enjoy the back alleyways and strange bars. We will have interesting experiences. We will drink a huge amount of alcohol. And we will publish a book with a front cover made of sandpaper.*

'Sounds spiffing!' chipped in Ralph.

'We will be truly free, enjoying life not in the limited ways others have decided for us, but in its infinite possible forms. It will be a movement of art, politics and philosophy all at once. And it starts . . .'

He paused again for effect.

'Now!'

Nothing happened.

Guy picked up his wine glass to signify he was finished talking, and looked directly at Piero. Piero held his gaze, then quietly lifted his hands and clapped them together once, twice, a third time. He looked around the table as his clap began to gather pace, but finding himself alone, opted to withdraw from applause at the eighth clap.

'Ralph, tell them about psychogeography,' demanded Guy, increasingly confident it was going well.

'Will doodly do! So . . . I've founded something called the London Psychogeographical Committee.'

'What's that?' piped up Elena, cheerily.

---

* They did, and it was.

'It's a mirage,' grinned Ralph. 'It's just me, I made it up. But because I made it up, it's real. See? Guy loves it.'

Guy shrugged.

'But what does psychogeography *mean*?' Elena insisted.

'It's about examining the specific effects of the geographical environment on the emotions and behaviour of individuals,' explained Guy.

'It's about how places make you feel,' simplified Ralph, to Guy's irritation. 'Anyway, the point is, we're going to produce psycho-geographical studies of cities – maps and things – that explore places as they've never been explored, but in the only way that truly matters: by mapping our subjective reactions to different places.' Then he added 'Hurrah!' and took a big swig of wine.

'Naturally, the maps we create will not be conventional,' expounded Guy, worried that Ralph might end up with all the credit for psychogeography, which he felt was his idea. 'Our maps will be anti-maps. The opposite of normal maps. We will not map the places people are conventionally interested in, but the hidden bits. And *obviously* we're not going round doing some dull capital-ist survey of the streets. No. The way we will create our maps, is through something I call the "*dérive*".'

'"The drift",' translated Ralph, for no one but himself.

'The *dérive* is a walk with no real beginning and no real destina-tion. The purpose is to let the city take you where it wants to, and to be guided only by your feelings towards the environment. This way, we'll resist the dull and traditional routes of the city, and instead create new and exciting situations along the way.'

'Didn't the Surrealists walk around like that?' asked Piero.

Asger Jorn, who'd been keen to contribute, now piped up: 'The Surrealists walked so that we could run – albeit, our running *also* takes the form of walking.'

'And our walking won't just be romantic, it will be *critical*,' stressed Guy.

'I'm not sure I follow,' admitted Walter.

'But how do we actually do a *dérive*, and make sure we're not just following the old predictable routes?' asked Michèle.

Guy swilled his wine while contemplating the question, which was admittedly a good one. Then Ralph had a thought.

'I once got lost in Cologne and couldn't ask the way since I didn't speak German. But using a map of London, I soon found the restaurant.'*

'I love that story,' exclaimed Guy.

'I'm sorry, what?' frowned Walter.

'We can use maps of different cities to explore our own,' proclaimed Guy.

'Navigate where?' continued Walter. 'Sorry, how can you find a restaurant that doesn't exist in the city with the map you're looking at?'

'I know,' explained Ralph. 'Perhaps, the restaurant was a mirage . . .'

Then he added: 'Isn't everything?'

At this point, Walter decided it was easier to pretend to understand.

Satisfied that everyone was now on board, Guy raised his glass and proposed a toast. 'To resisting the urban structures of oppression, both visible and invisible.'

The group raised their glasses, and the Situationists were born.

1.15 P.M., LONDON

Mark: Having navigated the locality of a large south London
    secondary school over lunch break, I've just picked up a
    lovely little coffee in a strange little place I'd never seen,
    where they were playing traditional Colombian music that
    rather effectively transported me to South America. Then
    I stepped outside again and was jarringly reminded of where
    I was and what I was doing.

---

* *This*, if nothing else, is a direct quote.

Jay: Exciting news, I must be close. I showed my map to a taxi driver, asking if he could take me to Place de la Concorde, where the obelisk is. Amazingly, he said he could take me to an obelisk. Turns out there's one just outside Euston station that I never knew existed. The disappointing thing is, I've obviously not been paying attention because he took me back the way I've just come from and charged me £8. Anyway, all this means that I must be extremely close. According to my map, all I have to do from here is find the longest, straightest road starting from the obelisk, and if I follow in a dead straight line, it'll lead straight to the Champs-Élysées. So that's exactly what I'm going to do.

Mark: I'm coming to the conclusion that London is either perishingly small or this experiment is doomed, because I keep walking down roads that I feel very familiar with. I decided to take a right, which, on my map, would have to be Rue Montmartre, to try to circle round towards the Champs-Élysées, and I've ended up walking past the apartment I rented when I first moved to London.

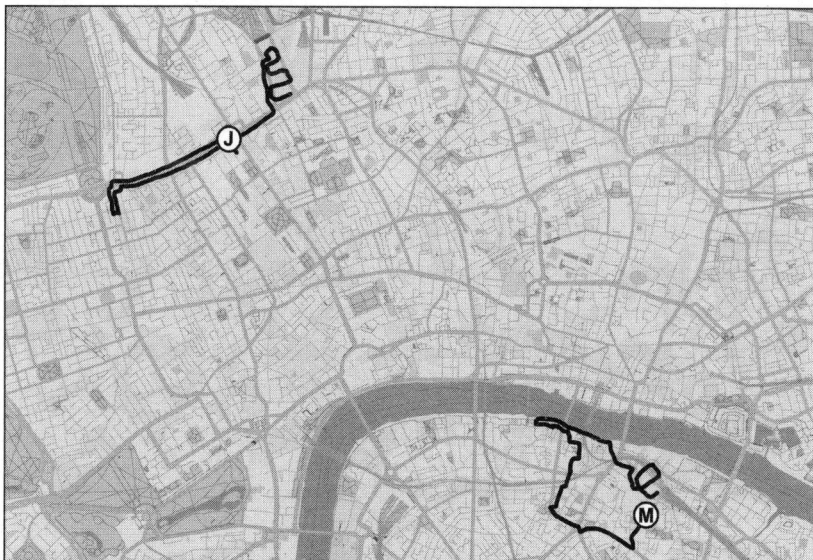

(*Map data from Open Street Map.*)

For a moment I thought I felt energised by some aggressive graffiti, but it was probably just the powerful Colombian coffee.

Jay: I've hit a brick wall. Literally. After about 30 seconds following my straight line, I've come across this building, which is not on the map and is in my way. While it does look like Victorian architecture, the only plausible explanation is it's a new-build that sprung up just after my map was published, made out of very old bricks. Hopefully, if I wiggle around it, the destination will be in sight.

THE NAKED CITY MAP

The Situationists believed that maps were a tool of capitalism, guiding people down repetitive routes to destinations deemed important by those with wealth and power. Using maps in bizarre and novel ways to subvert these expected routes became their way of revealing the map's inherent corruption.

In 1957 (the year of their founding) Guy published *A Psychogeographic Guide to Paris*, which included his most famous work: *The Naked City Map of Paris*.

This map shows 19 different parts of Paris cut up from a travel map and expressed in terms of their relationships to other places by means of large arrows. You might notice it's missing some things we'd normally expect to find on a good map: there's no scale, no legend, no grid and no clear geospatial relationship with actual Paris.

But this map wasn't concerned with the co-ordinates of places. Instead, it was supposed to represent a lived city, highlighting the emotional connections between places. Debord believed that:

cities have a psychogeographical relief, with constant currents, fixed points and vortexes which strongly discourage entry into or exit from certain zones.

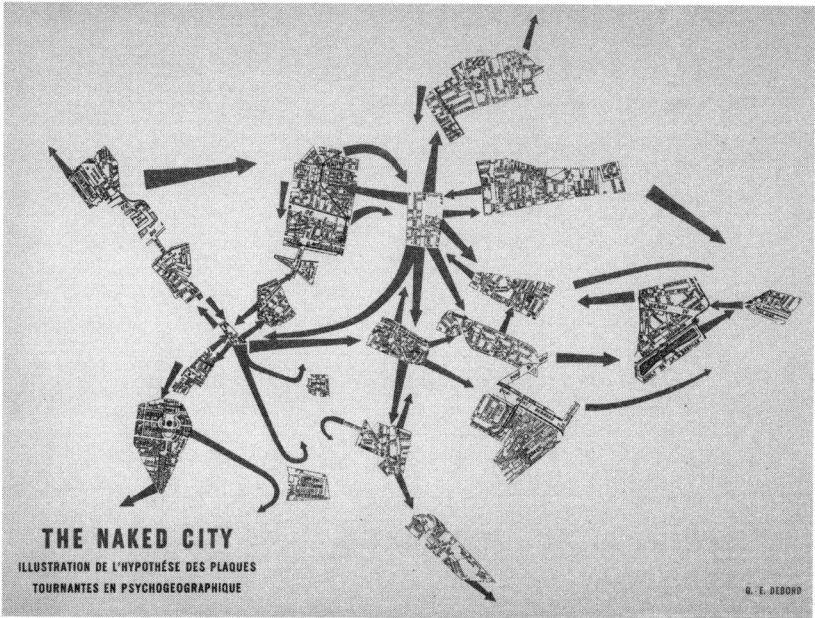

THE NAKED CITY
ILLUSTRATION DE L'HYPOTHÈSE DES PLAQUES
TOURNANTES EN PSYCHOGEOGRAPHIQUE

The Naked City was an attempt to map that. It might not help you actually find the area between Place de la Contrescarpe and Rue de l'Arbalète, but it will tell you that the same region is conducive:

> to atheism, to oblivion, and the disorientation of habitual influences

Which, in a way, might also be good to know.

As anti-capitalist protests go, there's no known record of this map forcing an emergency board meeting at Coca-Cola. But Debord's work, for all that it can at times appear quite baffling to us blind followers of capitalism, was, in fact, deeply influential, most immediately and profoundly in the famous May 1968 protests, when France was brought to a standstill in the most serious political unrest since the Second World War, almost toppling the de Gaulle government. Situationists were a central part of the protest, as workers and students occupied factories, universities and other institutions across the country. It was a moment when all the anti-capitalist sentiment espoused by the group seemed to take hold, and Situationist slogans were found graffitied across the country.

Slogans like:

Live without dead time

and:

Underneath the paving stones, the beach

Perhaps this could have been the springboard Guy's followers needed for them to go on and change the world. Instead, in typically revolutionary – perhaps nihilistic – Guy Debord fashion, it signalled the beginning of the end for the Situationists as an active movement. In 1972, he wrote:

Now that we can flatter ourselves on having acquired the most revolting celebrity status among this rabble [the left], we will become even more inaccessible and clandestine. The more famous our theses become, the more obscure we ourselves will be.

Which was maybe a tad frustrating for all those who wanted Situationism to *achieve* something.

1.50 P.M., LONDON

Mark: I found myself in a little garden with a small circular path going around a patch of green. I decided to walk around it three times to see if I noticed anything new each time I went around. Apart from the addition of a second crow on the patch of grass, I did not.

A man was sitting on some nearby steps eating a sandwich, watching me circle and making me feel self-conscious. He was wearing a beige top and beige trousers in a slightly different shade of beige – so I decided he should probably feel self-conscious too.

If I'm very honest, I don't feel remotely like I'm ten minutes away from the Champs-Élysées. If this *dérive* is also about mapping my emotions, then confusion is the one that is dominant.

I wonder how Jay is getting on.

Jay: I'm definitely on the Champs-Élysées! The map says I should be walking past the Jardins des Champs-Élysées on my right – and sure enough, there is indeed a long, thin, leafy park to my . . . left. Close enough.

I've come unexpectedly soon to what looks like a much larger park, not on the map, surrounded by iron gates with signs saying 'Russell Square'. This must be the 'reduced traffic and increased greenery' they've been promising for the famous boulevard for years. All I have to do to maintain my straight line is walk around the gates, turn back on myself,

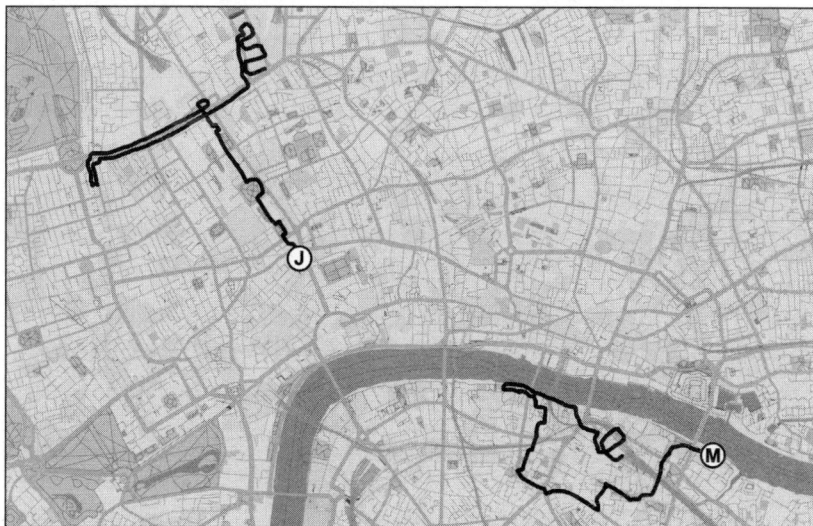

(*Map data from Open Street Map.*)

plough straight through the square, straight through the
fountain (might not bother with that), and then do the same
thing at the other end. I think I can see some restaurants on
the other side.

Could it happen? Could we be about to meet for lunch
after all?!

THERE'S AN APP FOR THAT

Guy Debord was undoubtedly a brilliant polemic thinker, whose
essays included clever-sounding sentences like:

> The introduction of alterations such as more or less arbitrarily
> transposing maps of two different regions, can contribute to
> clarifying certain wanderings that express not subordination to
> randomness but complete insubordination to habitual influences.*

But the truth was that he was an unapologetic alcoholic, and
most '*dérives*' were little more than glorified pub crawls.

By chance or design, the Situationists' academic fieldwork pur-
suits were incredibly well suited to regularly stopping at drinking
spots. They probably didn't need to navigate using a map of a different city to get inexorably lost and elicit new and dynamic 'situations', given the enormous amounts of alcohol some of them were
fabled to consume.

Guy himself was a passionate and volatile personality, appearing to revel in expelling people from the outfit. Less than a year
after its founding, Ralph Rumney had been kicked out, theoretically for the late submission of a psychogeography project from
Venice, and that was just the start. By the time the Situationists
were officially disbanded in 1972, 66 of the 70 members thought
to have passed through its ranks had either been expelled or

* *Parklife!*

resigned – the expulsions often receiving wonderfully melodramatic write-ups in the Situationists' own journal.

By this time, Debord had become more disaffected with the world than ever. He became less motivated and increasingly isolated. Health complications arising from a life of drinking, as well as his depression, have been cited as the reasons Debord took his own life in 1994, shooting himself in the heart.

But whether he would have liked it or not, his ideas have lived on, finding new forms and expressions in a world very different from the one he lived in 70 years ago. Banksy, for instance. What could be more Situationist than using art to give new expression and meaning to disused parts of the urban environment?

Psychogeography, too, has gone 'mainstream'* in recent years, not just as a subject of study, but also in conversations around urban planning, architecture and design. How we *feel* about places has become an important question in their development and construction.

Even the *dérive*, as a practice, lives on, and not just the one we genuinely did for this chapter. You can, for instance, download the *Dérive* app, which helps guide you through a randomised 'drift' of whatever city you're in. It gives you instructions like 'Follow a taxi for half a mile,' or 'Walk back two blocks,' or 'Hike to a public transport stop you've never been to.' At the time of writing, it has 3.8 stars on the Apple app store.

But would Guy himself have approved? Unequivocally no. If he'd thought we were living in a dystopian artificial reality called capitalism in the 1950s, one can scarcely begin to imagine what he would have thought about the role of the internet, mobile phones and Big Tech today.† 'The Spectacle' he described is surely more powerful than ever, our lives increasingly funnelled into a small

---

* For geography.

† A 21st-century *dérive* might now take the form of a drift through the internet, revealing hidden URLs among the 1.1 billion on the World Wide Web, intentionally avoiding the small handful that monopolise our online time.

handful of apps that claim to service our every human need while obscuring wider possibility, creativity, subjectivity, originality and everything else Debord held dear.

Perhaps the revolution is just around the corner. Perhaps we're reaching a saturation point just this minute. Maybe as we all start to notice the terrible reality of global phone addiction, more of us will take steps to escape the clutches of technology, consumerism, and never-ending accumulation by going completely 'off grid'.

Or, perhaps we won't.

Instead, we might consider embracing smaller, everyday acts of rebellion. Because there is something undeniably life-affirmingly novel about the simple act of turning off the blue dot, stepping into the big wide world, and submitting to getting completely and hopelessly lost . . . with or without a map of Paris.

2.12 P.M., NORTH AND SOUTH LONDON, PHONE CALL

Mark: Hello. So, we didn't meet.

Jay: Well, I think I found the Champs-Élysées but it's been renamed Sicilian Avenue and is much less impressive than I remember. I'm on my way home now.

Mark: You're not eating?

Jay: There *was* a Spaghetti House where I ended up, but it was closed for refurbishment. I've got cold spaghetti bolognese at home so I decided the map was telling me to go home and eat that.

Mark: I found the whole thing quite hard.

Jay: I really enjoyed it! I had new experiences, spoke to people I wouldn't have spoken to, saw bits of London I didn't know about, *and*, as a bonus, I feel like I know Paris much better as well.

Mark: I mostly found myself avoiding schoolchildren and walking up and down places I knew quite well. Now I'm eating a cheese toastie in a Starbucks. Guy Debord would not have approved.

Jay: Who?

Mark: I'm near Tower Bridge by the way.

Jay: We're not close. Sounds like you found it unsatisfying.

Mark: I think I wanted to get really lost, and I didn't. It doesn't help that I could almost always see the Shard. I think I like the idea of some of the things the Situationists talked about, but the reality of using a map of one city to navigate another was, in the end, as stupid as it sounds.

Jay: Well, we could always give it another go. How about next week we find our way to Legoland using dishwasher instructions? Mark? Helloo??

4

THE MAP THAT MADE
UP MOUNTAINS

'There are no foreign lands. It is the traveller
only who is foreign.'
Robert Louis Stevenson, 1880

THE SATURDAY CLUB

This story starts at 7.05 p.m., on 9 June 1788, in St Alban's Tavern, near Pall Mall in central London, on the first floor, in the private room at the back, at the long table by the back window.

This is where, every Saturday, Sir Joseph Banks – naturalist, botanist and president of the Royal Society – used to attend a gathering of London's upper-class establishment known as the Saturday Club for a spot of intelligent company and conversation. Only the very wealthiest and very most influential men, with the whitest, fluffiest wigs, were allowed to be members.

On this particular Saturday, Joseph and his friend Henry Beaufoy MP were having a good stare at a map on the wall. Pubs in those days – just like today – often had decorative old maps on the wall, except it *was* the olden days so the maps weren't that old yet. Admittedly, we don't know for sure that this map was literally on the wall, but Joseph was certainly aware of this map, and for the purposes of storytelling it suits us if it was on the wall in the room where this bit of the story takes place.

The map in question was a map of Africa by French cartographer Jean-Baptiste Bourguignon d'Anville.* It was the best and most accurate map of Africa available in Europe at the time (which by now was 7.06 p.m.), though that's not really saying much.

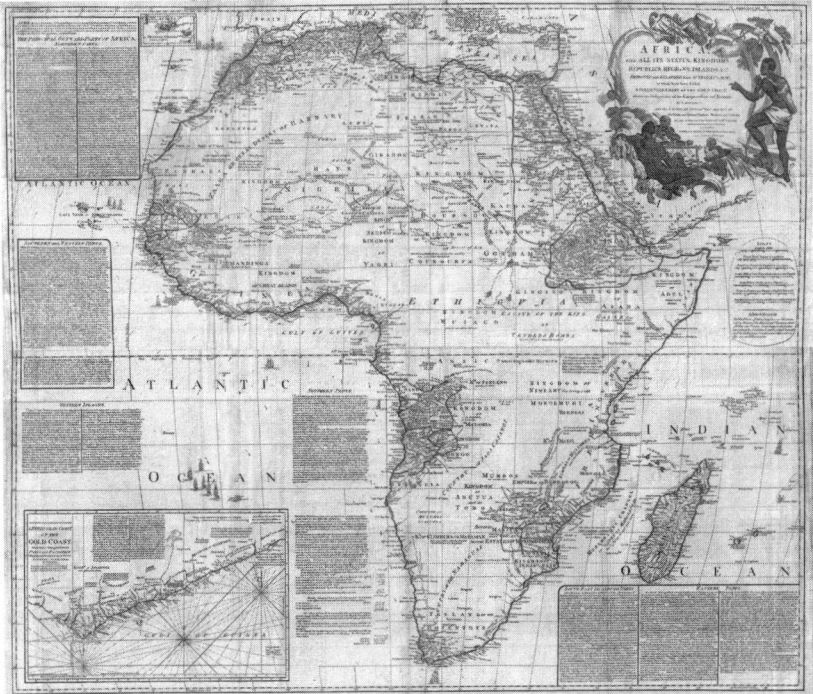

'What an awfully dull place,' said Henry Beaufoy. 'Nothing but vast tracts of bleak and featureless nothingness across thousands of miles. Is it any wonder that everybody calls Africa the "Dark Continent"?'

'Alas no, Henry,' said Sir Joseph, shaking his head. 'It has that name, and the map has this form, because we know so little about it. It is a great failing of the Age of Enlightenment that, in a time when men can sail around the world, the geography of Africa remains almost entirely uncharted.'

---

* Translated and slightly embellished by British cartographer Samuel Boulton. But d'Anville did all the difficult bits.

Sir Joseph had a good point. In the 18th century the only reliable knowledge Europeans had of the African continent was its overall shape, which cartographers were very familiar with from centuries of trading up and down the coast in gold, spices, textiles and ivory.

And, of course, slaves. For European countries with navies, the African coast represented an opportunity to make themselves richer and more powerful by whatever means necessary, including – indeed, in particular – off the back of acute human suffering. You can't tell a story about how Europeans came to understand Africa without also addressing the fact that one of their key vested interests was the exploitation of the people living there in one of the darkest and most shameful chapters in recent human history. Well, you can, but that would be like pretending it didn't happen, which is much worse.

In the 18th century this trade was still largely confined to the coast, which explains the impressive, even surprising accuracy of

We weren't making up how horrible it really was.

the coastline on d'Anville's map, in contrast to the gaping vast-ness of blank space in the interior, radiating an almost audible Gallic shrug. And what scarce little detail there is, at best is com-pletely wrong, at worst downright offensive. The sparser the geo-graphic features, the more outlandish and horrendously racist – even for those days – the annotations become. Lowlights include 'a dwarfish breed who hunt Elephants', 'always at war' and 'Men-eaters'. It was also very much in vogue at the time for Euro-pean maps of Africa to include absurd flights of fancy, like myth-ical one-eyed monsters.

The problem for Europeans was that the African interior was very hard to explore. First, the terrain was challenging, with sand-swept deserts and dense rainforests. Second, diseases against which Europeans had no protection – such as malaria and yellow fever – were rife. Third, the climate was unfamiliar and unforgiv-ing (very hot). And fourth, the locals often made it violently clear that they weren't too happy about having visitors with the sorts of intentions this lot seemed to have.

As a result, European mapmakers in the 18th century had to work with what limited information about the African interior was available, and information about the interior came from some pretty suspect places. One commonly used source was writ-ings by ancient Greek and Roman authors such as Pliny the Elder and Ptolemy. Although they were pioneers of geography, mathematics and astronomy in their day, that 'day' was now an awfully large number of other days ago. Their accounts of the African interior were more detailed than anything any European had come up with in more than a thousand years, but they were less than ideal for 18th-century mapmaking. This was partly because they were massively out of date – entire empires had come and gone from the continent in that time – and partly because the ancients didn't actually use scale maps. In those days, the geography of places was recorded with descriptions, like diaries or travel guides, a method that's practically begging for misinterpretation and mistakes.

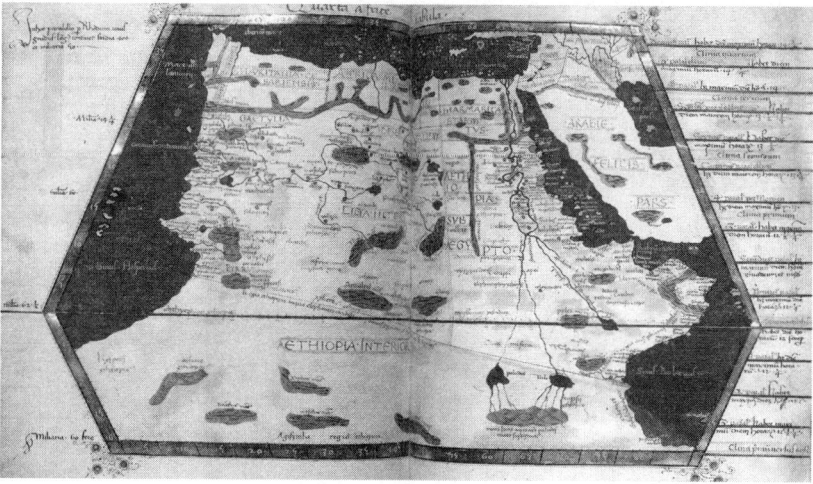

This is a map of North Africa from 1467 using Ptolemy's notes from more than a thousand years earlier. By today's standards it's rubbish.

Another source was a slightly, but not adequately, more recent account of the African interior from the 16th century by the famous traveller Leo Africanus.* Leo, who was born in what was then Muslim Andalucia, travelled all over Africa as a diplomat before being captured by pirates, taken to Rome and made to convert to Christianity, which made him uniquely placed to provide first-hand insights for Europeans into Africa's geography, cultures and trade. His diaries were very popular and widely known in the 16th century, especially the bits with juicy details about the semi-mythical city of Timbuktu.†

So that's why d'Anville's map in the pub was so awful (we should be clear that d'Anville's wrong map isn't the wrong map of this chapter – that's yet to come, and it's worse); he'd had to fill in

---

* This was a nickname he acquired because he knew so much about Africa. His real name was al-Ḥasan ibn Muḥammad ibn Aḥmad al-Wazzān al-Zayyātī al-Fasī.

† Timbuktu isn't mythical at all. It's very real, and you can go there right now if you like. But in those days very few Europeans had been there, and many believed the entire city was made of gold, which was only 0.00002 per cent true.

the interior using unreliable, unrelated and unreconcilable source material. As a result of this, one of the things d'Anville's map was geographically wrongest about was one of the longest and most important rivers in the continent – the River Niger.

Europeans knew from various sources and interactions with merchants on the West African coast that south of the Sahara Desert there was a mighty river connecting Timbuktu to the sea. What they didn't know was which bit of the sea it flowed into, how wide it was, where its origin was, which direction it flowed in or how to pronounce it.* Every source seemed to suggest it did something different, with some saying it went west into the Atlantic, others that it went east and joined the Nile. On d'Anville's map, the River Niger squiggles eastwards and then stops in the middle of the desert, which rivers generally don't do.

Sir Joseph tapped repeatedly on the glass protecting the map with his fingernail. '*That's* the bit that bothers me the most. The River Niger.'

'What about it?' said the Bishop of Llandaff, who'd stood up from a nearby table and sidled over to join in.

'That river,' said Sir Joseph, still tapping, 'is the key to the whole continent. Where there's a river, there are people. Farmland is irrigated, people are fed, and long-distance travel and trade are enabled. To know the River Niger is to know the African interior.'

Then Henry Beaufoy piped up, 'Well, why don't we find the River Niger ourselves then?'

Sir Joseph stopped tapping. 'Ohh . . .?' he said, turning to Henry and raising one eyebrow disconcertingly slowly.

'We are men of science, are we not? What could be a more worthy scientific pursuit than uncovering the mysteries of the mysterious continent?' Beaufoy had caught the attention of the other members, who were all now gathered in front of the map.

* 'nee-ZHAIR' or 'NIGH-jurr'. Either one's fine.

'We could collect specimens of new plants and animals,' said Lord Rawdon.

'Or discover valuable minerals and resources,' said Sir William Fordyce.

'And better trade routes for gold and ivory,' said Sir John Sinclair.

'We could find millions more customers to buy British Things,' said the Earl of Galloway.

'Better knowledge of the peoples of Africa will shift public opinion towards the abolition of slavery,' said famous abolitionist William Wilberforce, which got a small applause.

'We could buy more slaves!' said Sir William Young, which got a bigger applause.

'We could mix lager and cider and put a dash of blackcurrant in it,' said Bryan Edwards, who hadn't really been listening.

'So, we're all agreed,' announced Sir Joseph excitedly. 'We're all *really* interested in finding the River Niger, and knowing Africa better in general. All in favour of making it a reality and doing it ourselves?'

'Aye!' they all shouted in unsatisfying drunken semi-unison.

'Knowing Africa better in general' was such a broad and vague aim, it was something that – despite their wildly differing, even conflicting motives – they could *all* get behind. And so, the Association for Promoting the Discovery of the Interior Parts of Africa – known for short as the 'African Association' – was born.

All members wealthily agreed to pay five guineas per year to pay for the recruitment and equipping of eager explorers. But despite the group's enthusiasm, popularity and money, it turned out sending a man to the River Niger was *really* hard. Explorer after explorer failed in their attempt to find the elusive river.

The first to have a go was John Ledyard in 1788, who tried approaching it from the east. He got as far as Cairo before coming down with a bilious complaint and trying to cure himself by poisoning himself to death with sulphuric acid.

Next was Simon Lucas, later in 1788, who tried to sneak up on the river from the north, but his route was blocked by warring tribes. His guides abandoned him and he limped home without ever getting any further than the coast.

Then, in 1790, Daniel Houghton tried approaching it from the west, but he was robbed and either abandoned or deliberately murdered by his guides or local tribespeople. No one knows exactly what happened to him, but he's definitely dead now.

It would take until 1795 for the Association to finally find somebody who would actually manage to reach the River Niger and live to tell the tale. And it was a very unlikely somebody indeed.

MUNGO PARK

Mungo Park's main area of expertise was botany. A young* doctor and devout Calvinist from rural Scotland, he was shy, bookish and had limited experience of exploring, but he had shown an interest in West Africa – possibly by filling in the wrong form. Sir Joseph Banks, who knew Mungo as a surgeon's mate with the British East India Company, thought he'd be great for the job, partly because, by this point, the Association was starting to run out of explorers. So he recommended that Mungo Park follow where Daniel Houghton had left off.

Mungo was by all accounts a kind, softly spoken and naive man, who seemingly never had a sense of any danger he was in. We know this because of the extensive diaries he kept on his travels.

His published diary starts on 22 May 1795, when he departs on a boat from Portsmouth.

But!

Our research has uncovered an as-yet-unpublished chapter that was written the day before. We're excited to be able to reproduce it exclusively here so you can get a flavour of Mungo's brave stoicism

* 23.

in the face of constant bad luck, as well as his tone and fine writing style, without having to read his entire book.[*]

So without further ado, it is a great honour to present . . .

TRAVELS IN THE INTERIOR DISTRICTS OF AFRICA BY DR MUNGO PARK

Chapter Zero

21 May 1795

Upon stepping forth from the front door of the Association of Africa's office in London, bound for Portsmouth, I was splashed immediately by a passing horse. The mire drenched my coat and breeches, and penetrated my eyes and mouth. While searching my pockets for a handkerchief, an unexpectedly icy gust of wind did congeal the mud into the whites of my eyes before I could cleanse them.

Temporarily blinded, I managed, with some difficulty, to beckon a small boy and ask him for directions to Waterloo station. The boy informed me in the local cockney dialect, which I had spent the prior weeks studying, that the steam locomotive wouldn't be invented for another nine years, nor Waterloo station for another 53. I thanked him for his kindness and offered him my shoes as a parting gift. He declined them, but I left them in the street for him to pick up later in case he changed his mind.

Barefoot, I ventured forth beneath a pelting rain, feeling my way by flailing my arms and legs before me, shouting 'Excuse me' repeatedly and very loudly. After four hours of such toil, I tumbled stiffly into the majestic waters of the River Thames, its waters thick with the city's refuse. Yet, by God's mercy, this foul tide did

---

[*] If you're reading it out loud, we suggest you do it in the voice of James McAvoy *in The Last King of Scotland* – a film about another young Scottish doctor going to Africa.

scour the frozen mud from mine eyes, and sight was restored unto me. 'Oh joy! Thank you!' I shouted to the heavens, as a leech did enter my mouth.

As dusk descended, I clambered up onto the north bank of the river, curled myself up into a tiny ball beneath a wooden pier and camped for the night. At dawn, I awoke to find all my clothes and belongings stolen, the chilly air bestowing my skin with a sweet freshness, giving me courage to continue my journey. 'Praise be!' I bellowed to my creator, as a family of leeches tumbled out from my lower lip, well sated.

A passing undertaker took pity on me. The driver was bound for Plymouth, which he informed me was only three days' walk from Portsmouth. He offered me passage, on condition I share his cart with a freight of his delivery of gangrenous corpses, and that I lie at the bottom, allowing further deliveries to be stacked atop over the course of our journey.

My heart leaps with joy to know that I am progressing towards Portsmouth, and with rich snacking provisions on hand to sustain me through the voyage.

If my unceasing good fortune continues in this manner, I might even procure myself some new clothes.

Thankfully for Mungo this optimism helped see him through the remaining horrendous (and actually true) misfortunes published in his real diary once he reached the continent, including:

- Catching malaria
- Getting arrested for not paying trader's dues and paying them off with half his belongings
- Befriending a local guide who took half his belongings again
- Getting caught between two warring nations in the desert
- Being robbed by bandits
- Getting captured and detained in a small hut he shared with a pig for three months

- Being robbed again
- Being detained again
- Being put on display and made to eat raw eggs to entertain the crowd
- Losing his horse
- Catching one more fever before making his way home to England on a slave ship (via the Americas)

In better news, he did successfully become the first European to set eyes on the River Niger.

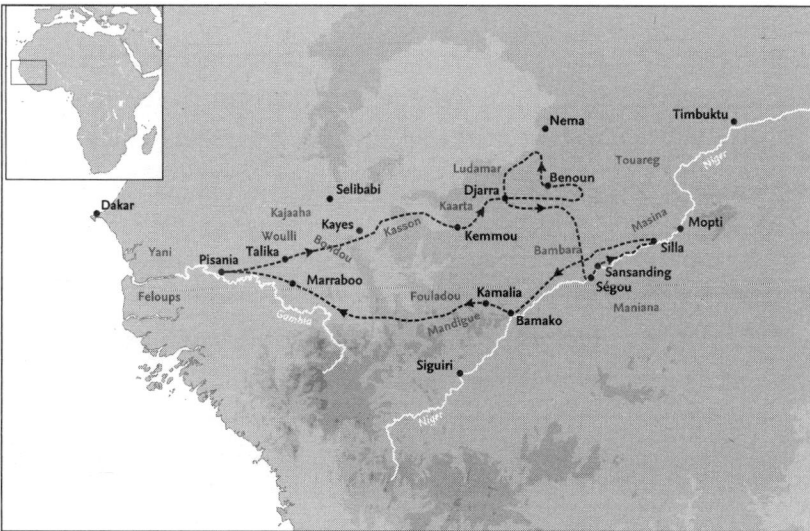

The route of Mungo Park's expedition through West Africa in 1795.

On 22 December 1797, when Mungo stumbled through the doors of the African Association's office looking very dishevelled, covered in mosquito bites, then raised a finger and weakly exclaimed 'I'm baaaaack,' the other members were flabbergasted. They'd grown rather used to never seeing their explorers again, so his appearance out of the blue came as a bit of a shock. He was treated to a hero's welcome: twelve pints of snakebite and fifteen verses of 'For He's a Jolly Good Fellow'.

Although he hadn't succeeded in following the River Niger all the way to the sea,* the members of the Association were thrilled by his achievement. He had made the crucial discovery that the Niger counterintuitively flows *away* from the nearest coast to the east before swinging north, something that no European had predicted.

Mungo's diary became a smash hit, not just in England but all over Europe. It was an incredibly influential book, correcting a lot of European misconceptions about Africa, and offering a more humanised view of African society, heralding an era of European fascination with Africa. It also made a hero of its author – the brave explorer, daring to go to dangerous, never-before-seen places for the sole purpose of finding out what was there.

Back in St Alban's Tavern, Sir Joseph Banks and the other members stood proudly staring at the map on the wall. 'Well, gentlemen,' said Joseph, 'we did it. Our determination and our money have finally brought greater knowledge of this fascinating continent.'

'Why are we still staring at this rubbish old map then?' said Sir Francis Burdett.

'Oh yeah, I forgot. We need a new map now, don't we?'

And so, the Association's next step was to commission a new thoroughly detailed map of the continent inspired by Mungo's thoroughly detailed book. And there was no member more perfect for the job than a man called James Rennell.

JAMES RENNELL

James Rennell was an English cartographer who'd cut his cartographic teeth as the Surveyor General of Bengal, making

* In case you're interested, the River Niger actually flows all the way through what's now Guinea, Mali, Niger (named after the Niger) and Benin, before entering the sea at the Niger Delta in what's now Nigeria (named after the Niger). Pretty circuitous route, but rivers aren't fussed about saving time.

impressively detailed and highly trusted maps not only of the
Indian subcontinent, but of all the sea on the way there and back,
earning himself the nickname the 'father of oceanography'. If
anyone could be trusted to use their geographical expertise* to
convert Mungo's memories into a proper map, it was him.

And so James set to work. But there was one problem. James
was a geographical perfectionist.† If there was one thing he
couldn't stand, it was blank space.‡ He wanted his map of West
Africa to look as full and scientific and as complete as geographically possible. He therefore endeavoured to fill it *all* in.

Mungo's diary had helpfully provided extensive details about
the landscapes, rivers, cities and kingdoms he'd encountered along
the way – all very useful stuff. But James's attention was particularly drawn to a passage from about three-quarters of the way
through the book. It appears during the bit where he's trudging
home – suffering from fever and exhaustion – along the banks of
the River Niger that he'd just 'discovered'. The passage goes:

> I gained the summit of a hill, from whence I had an extensive
> view of the country. Towards the south-east, appeared some
> very distant mountains, which I had formerly seen from an emi-
> nence near Marraboo . . .

'Mountains, you say?' muttered James, as he dipped his pen
purposefully into the ink box. So he began sketching some mountains on the canvas, and then a thought occurred to him. The
whole point of this exercise was to find the *source* of the River
Niger. If Mungo's description of the river's width, course and
bendiness in this part of Africa were true, then these mountains
he described must surely be the river's source. And if the river were

* geographexpertise.
† geographectionist.
‡ If there were two things he couldn't stand, they were blank space and
liquorice.

as wide as Mungo had described, these mountains would surely
have to be part of a much larger mountain range that Mungo
couldn't see.

And so James drew another little mountain just to the right of
his first one. And then another. And another, and another, and
kept going right across the page. Possibly to highlight the hypothetical nature of this mountain range, he drew them in an absurd,
geographically improbable single file, stretching halfway across
the entire continent.

Next, this mountain range needed a name. Standard practice at
the time was to name features after nearby settlements, rivers or
prominent individuals, especially when the locals didn't have a
name for them (which, in this case, they *obviously* didn't). So, to
pick a name, Rennell turned to the continuation of the very same
passage in Park's diary:

> . . . where the people informed me, that these mountains were
> situated in a large and powerful kingdom called Kong . . .

'Kong!' sang James loudly. And then, in his best handwriting,
he scrawled the now-infamous first-ever use of the words 'Mountains of Kong'.

James Rennell's 1798 map containing the first appearance
of the Mountains of Kong.

Importantly, James knew full well that the existence of these mountains was a guess, and that there was a good chance they weren't there at all. But his style of mapping didn't have a symbol for 'hypothetical'. So trusted was his authority on geography, and so well drawn were the mountains,[*] that the Mountains of Kong were assumed by everybody who gazed at them to be completely real.

Yet it wasn't just Rennell's trustworthiness that left these hypothetical hills open to misinterpretation. There had been a general change in map fashion around this time, with maps increasingly being no longer decorative but presented as enlightened, rational, scientific products. One-eyed monsters had fallen out of favour, and nowadays, if something was on a map, it was, quite simply, supposed to be there. The danger, however, was that at just the time when people expected their maps to be as true as possible, the legacy of pure invention had not quite been fully shaken off by all cartographers. This created the most dangerous kind of invention – the one that people assume to be true.

And things were only about to get worse.

'Gentlemen . . .' began Sir Joseph Banks, 'I present to you the fruits of your donations: a vastly improved map of Africa.'

Sir Joseph tugged on a cord, and a pair of dark red velvet curtains parted to show James Rennell's masterpiece, significantly taking the place on the wall where d'Anville's map once hung. Again, we can't know for sure that this is exactly how it happened, but it's a symbolic moment in the spirit of good storytelling.

'Phwoooar, look at that mountain range,' said Henry Beaufoy.

'A thousand miles from end to end,' mused Sir John Coxe Hippisley.

'The longest mountain range on the planet,' whispered Quentin Lazerquest.[†]

---

[*] A few simple lines with cross-hatching creating a 3D shadow effect. Looks really addictive to doodle.

[†] Membership of the Association had grown to more than a hundred by this point, with the names of many of the members lost to history.

Rennell's map was notable as the first map to show the Mountains of Kong, but crucially it was not the last.

In 1802, four years after his map appeared, another cartographer, Aaron Arrowsmith*, created the boringly but appropriately named 'Africa Atlas'. Trusting the authority of James Rennell's work, Aaron faithfully copied the Mountains of Kong. And Aaron Arrowsmith's 'African Atlas' would go on to be even more widely spread than Rennell's.

So now the Mountains of Kong were on *two* maps. And this, for all future armchair cartographers, made them an irrefutable feature just like the Sahara Desert or the Mediterranean Sea.

Between 1798 and 1892, the Mountains of Kong were a feature on practically *all* European maps of Africa. Historians today are aware of at least 40.

Here's one:

* Slightly silly name, but this one's real.

Here's another:

Although the mountains became a constant feature, they were never a consistent one. Over the decades, almost everything about the mountains kept changing. Some writers said they were 'barren', while others claimed they were 'snow covered'. Some said they were limestone, others granite. And, depending who you asked, they varied massively in height, from as low as 2,500 feet to as high as 14,000 feet. Their position kept shifting too, with some as far north as Timbuktu, and others as far south as the Guinea Highlands about 600 miles away.

The length of the mountain range varied massively too, seemingly getting longer as time went by, eventually reaching all the way across the entire continent, connecting up to the Mountains of the Moon on the eastern side of Africa (where the Nile was said to have its source), which, by the way, were also fictitious.

Despite the Mountains of Kong being imaginary, they nonetheless had a huge influence on how Europeans thought about Africa. Explorers steered clear of their vicinity, taking a completely unnecessary long way round on expeditions. The mountains were seen as an impassable barrier, preventing any thought of trade or travel between the two sides. Take a look at this map, for example. Here,

the Mountains of Kong are used as the border between Upper
Guinea and Soudan.*

'Map of Africa' plate in Samuel Augustus Mitchell's School and
Family Geography (1839).

You couldn't really ask for a clearer visual metaphor for how
Europeans thoughtlessly imposed the concept of national borders
onto a continent they'd demonstrably never visited. It betrays an
extraordinarily flagrant disregard for the people who actually lived
there, as if the map of the place *were* the place itself.

But, of course, the Mountains of Kong inevitably had a shelf
life and they couldn't stay on maps forever. It was only a matter
of time before somebody happened to be the first person to *not*
see them.

In the late 19th century Louis-Gustave Binger, a French explorer
with a medium-sized moustache, irritated by the mountains' incon-
sistency, wanted to find out for himself how tall/rocky/snowy they
were once and for all. So in 1887 he set out on an expedition starting in Mali, following the River Niger upstream. After two years
of trekking he was unable to locate any mountains at all. Frequent
stopping and asking for directions led him to find out that while

---

* Spelt wrongly.

there was a city called Kong, none of the locals knew anything about any mountains.

At first he was disappointed, especially having schlepped his skiing gear all that way. But then it dawned on him that he'd found something much better than massive mountains – a massive scoop.

When he got back to France in 1889 he told everyone that the Mountains of Kong had been fake all along and he had maps with far more detail than ever before to prove it. From the 1890s the mountains began to quietly disappear from everyone else's maps, replaced by what's really there – mostly flat or gently rolling terrain, with a few lumps now and again, but nothing to write home about mountainwise.

It's a sad yet important fact that this story doesn't have a happy ending. The true geography of Africa finally being understood by Europeans did not lead to an age of enlightenment and harmony between the two continents, but instead enabled and encouraged European powers to divide the continent up for themselves and exploit its resources. In 1893 Louis-Gustave Binger went on to fulfil the ultimate purpose of his fact-finding, map-correcting mission: the Ivory Coast became a French colony, with himself as governor, all part of the so-called 'Scramble for Africa', which reached its apogee at this time.

So what became of the African Association? Surprisingly, it was never officially disbanded; its members and its aims were absorbed by the Royal Geographical Society in 1831, which means it *technically* still exists, but only in the meaningless theoretical way that Yorkshire Television does (see next chapter).

As for Mungo Park, ever the glutton for punishment, he went back to Africa in 1805 to finish the job he started. But despite managing to travel much further the second time, he still didn't find the end of the River Niger, or Timbuktu, and arguably had even worse luck than before; his boat got stuck in rocks at the rapids of Boussa, and while hostile natives threw spears at him he was forced to jump into the water and drown.

Mungo's reputation today is somewhat mixed. His name comes up everywhere you read about the Mountains of Kong, history having decided they were all his fault. Nonetheless, there's a statue of him in his native Selkirk.* Had he lived another 200 years to google himself and see what had become of his legacy, he'd no doubt have been at pains to point out that he never mentioned a long mountain range in his diary at all, let alone a comically improbably long mountain range.

So, should James Rennell get the blame instead?

Satisfying though it would be to pin the blame squarely on one individual's ignorance, incompetence or hubris, as with most of real life, the truth is more complicated and doesn't follow the rules of good storytelling. The Mountains of Kong were a group effort. It took a combination of Park making a big deal out of a hill he saw, Rennell doing his best with limited information to finish what he thought was his job and subsequent mapmakers making maps based on the best maps. None of these on their own is a serious crime, but when their powers combine . . . Boom! Fake mountains.

Today, the Mountains of Kong are a curiosity, present on numerous old maps over a period spanning nearly a hundred years – a symptom of an embarrassing period of European history and mapmaking that spanned the 18th and 19th centuries. They serve as a towering phantom monument to European ignorance in protocolonial times, and are a foreboding sign of a much darker era to follow.

Surprisingly, the mountains didn't disappear entirely. Somewhat unbelievably, they managed to make an appearance in *Goode's World Atlas* as recently as 1995, although it's likely this was just a copyright trap[†] or an in-joke for map geeks.

---

* They were going to name the town's park after him, but they'd have had to have called it Mungo Park Park.

† These are fascinating, by the way, but now's not the time. Chapter 7 is the time. Go read Chapter 7 (after Chapters 5 and 6).

While the Mountains of Kong might be the wrongest, they were far from the only false feature that made their way onto otherwise believable maps around this time.

In 1827, for example, cartographer Thomas J. Maslen produced a map of Australia for his book *The Friend of Australia*, showing, among many other completely guessed inaccurate features, a ginormous lake, bigger than Belgium, slightly to the right of where Uluru should be.

In a remarkably similar story to the one that gave us the Mountains of Kong, the Australian continent's outline was well known, but its interior wasn't. Rivers near the east coast were found to be flowing counterintuitively away from the ocean, as if pulled by a larger, as yet undiscovered body of water. Thomas therefore added a hypothetical 'Delta of Australia' to his map, including a ludicrously large lake that wasn't there.

Incidentally, years later when it was eventually discovered that the only lake in this area was far smaller than expected, it was

given the very appropriate and superbly Australian name 'Lake Disappointment'.

What makes maps like these ones so striking is the contradiction of an impressively accurate coastline with a hilariously inaccurate middle. A map from many hundreds of years ago that's entirely rubbish, with borders drawn seemingly by a five-year-old, is not very interesting to look at. Neither is a modern, accurate one. But in this very specific sweet spot in the late 18th and early 19th century where these two contradictory mapping styles very briefly intersected, we're treated to some real gems like the Mountains of Kong and the Delta of Australia, that are endless fun to stare at and laugh at.

Today, of course, speculative cartography is no longer a thing, and it's reassuring to know that in the 2020s all maps of Africa, and indeed everywhere else, are perfect and there are no inaccuracies on maps ever.

5

THE FUZZY MAP

*'It is better to fight and lose than to
have fought and not won.'*[*]
Ethelbert Talbot, 1903

While Britain can pride itself on a lot of things – very polite
drivers, the world's most robust plug sockets and an unparalleled
variety of crisp flavours – one thing it's never been terribly good at
is splitting itself up into bits.

For comparison, let's have a look at France just next door. France
is divided neatly into 13 regions, which are further subdivided into
96 departments.[†] These play a pivotal and visible role in many
aspects of French life, from local government to postal addresses
to car numberplates, with all the systems working consistently and
harmoniously. The effect of this is that every French person knows
where they are and which of their neighbours they band with for
various regional identity purposes. Put simply, every French person
knows their place.

But ask a Brit what part of the country they're in, and there are
a hundred different ways they could answer your question.

There are dozens of different systems the UK uses to divide
itself, each for a different purpose and each using a different

---

[*] Nothing to do with this chapter, we just thought it was a nice quote.
[†] Technically 18 regions and 101 departments if you count the Overseas
Territories.

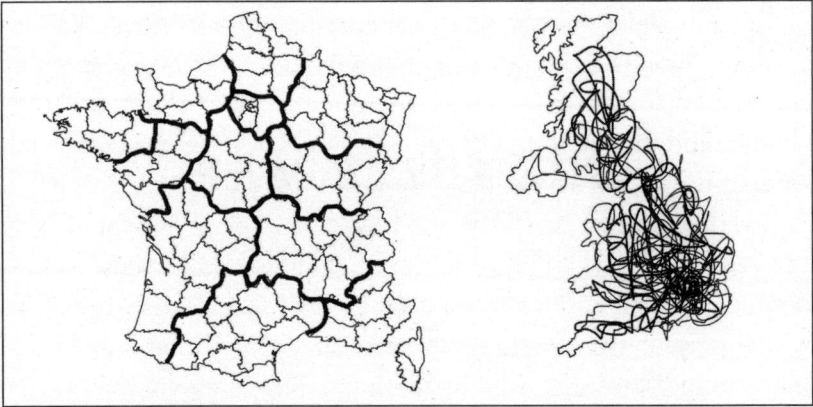

France: neat and tidy. Britain: all over the shops!

map, with borders that tangle in and out of each other like head-phone cables.

For starters, there are counties – the historic subdivisions dating back to the Saxons, mostly ending in 'shire'. These come in wildly varying sizes, and even more wildly varying degrees of local pride. They're not a very helpful distinction for identifying where you live, because competing with the historic counties is the infuriatingly similar system of administrative counties for local government, which uses a slightly different map. Residents of Saddleworth for example can't, under pressure, say whether they're in Greater Manchester or Yorkshire. This isn't helped by yet a third definition of 'counties' used in British addresses. Royal Mail uses their own map to avoid splitting postal towns that straddle traditional county borders. This puts, for instance, the Essex town of Ugly Green in 'Hertfordshire'. And these incidentally are entirely unrelated to the Post Office's postcodes – a system initially set up in the 1960s for the benefit of postal workers, but which has since become indispensable for sat navs, emergency services, estate agents and rival street gangs. And these postcodes don't match up at all with phone codes.

There are parliamentary constituencies that don't match up with local government areas, NHS areas that don't match up with police areas, and so on and so on.

None of these systems feels satisfactorily authoritative. And, as a result, any nationwide organisation that needs to carve the country up for whatever purpose can do so any which way they choose, adding yet another layer of complexity and confusion. It's something we Brits have just grown used to.

But there exists one particular jigsaw of Britain that for most of the second half of the last century made very little sense indeed despite playing an enormous role in how the British identified which part of the country was theirs. The people who drew this map caused considerable confusion and controversy, some of its effects lasting to this day.

We are talking about the map used by ITV – the BBC's main commercial competitor in the UK – to split the UK up into different regions, each receiving their own local television service with local news and local programmes.

The eagle-eyed among us will have spotted that some bizarre decisions have been made here. According to ITV, Wales spills over into Bristol, part of Norfolk is in Yorkshire, some of Scotland is in England (or vice versa), and there appear to be two Londons. So what on earth happened here?

To find out, we need to take a dorky deep dive into the nation's nebulous and nuanced network infrastructure.

Strap in.

THE BLACK AND WHITE CHANNEL

Before 1955 the BBC (British Broadcasting Corporation to its friends) provided the nation's only television service. If you were rich enough to own a television set, which back then was a heavy wooden cupboard with a screen the size of a dinner plate that made a high-pitched whining noise, your annual television licence funded the BBC, who beamed out all the programmes from their London headquarters in Alexandra Palace.

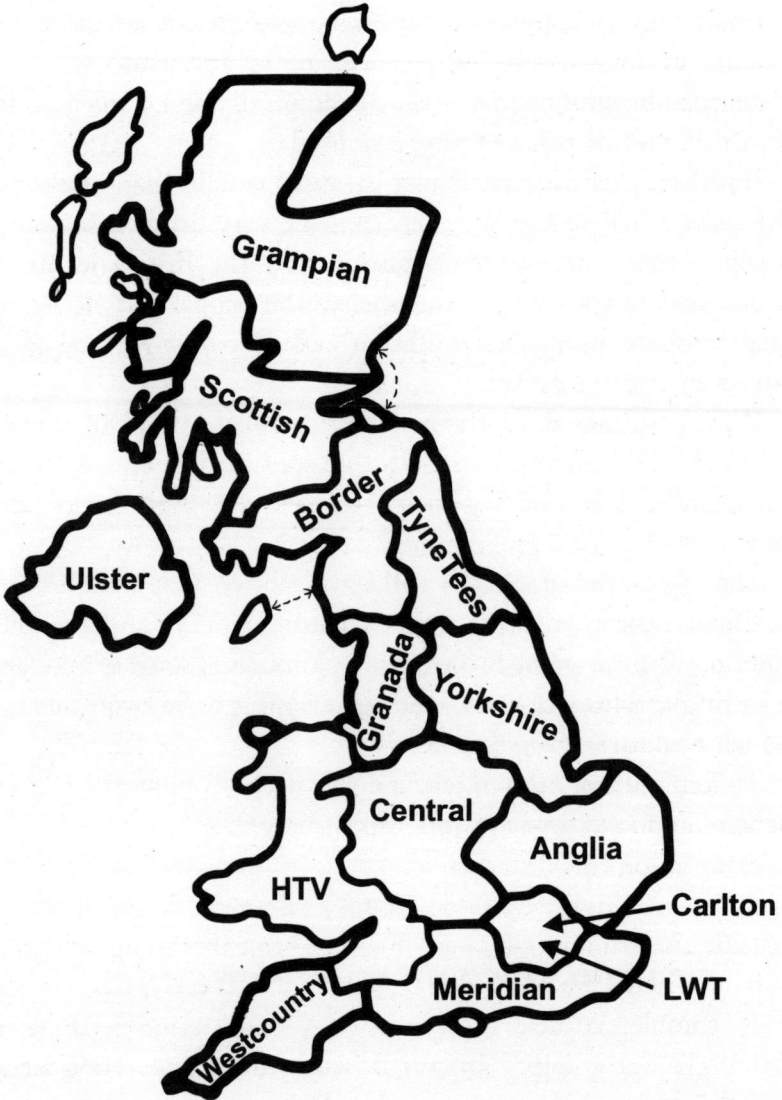

ITV regions. If we're being specific, ITV regions between 1993 and 2006, when this map was at its worst.

In those days there was no concept of 'changing the channel'. If you didn't like what was on, your options were to boo loudly until the programme was over, switch it off and listen to the wireless, or go outside in the garden and dig a hole while singing the National Anthem. But as the 1940s turned into the 1950s, and television sets became increasingly popular, Brits with tellies were starting to ask, 'Shouldn't we have more choice?'

After much debate* in parliament, in 1954 the government announced that commercial television was coming to Britain. The Beeb's monopoly was to end and viewers would finally be given 'freedom of the knob'.†

This news caused a lot of moustaches to nervously bristle over breakfast. At that time, the only other place in the world where commercial television existed was in the United States. The American way of doing it was, to a British sensibility, positively vulgar.

On American 'TV' (an abbreviation we've successfully avoided until now), shows were individually sponsored. The advertisers had a lot of control over the content of the shows they were attached to, which meant the programme and the sponsor were often blended together. Halfway through *The Flintstones*, Fred might turn to the camera and say, 'Sorry to interrupt, Wilma, but I just wanted to take a puff on this flavorful Winston's cigarette,' or if a contestant got an answer right on *You Bet Your Life*, Groucho Marx would run outside the studio and smash his way back in through the wall driving a new Chrysler.‡

If commercial television insisted on becoming a thing in Britain, it was going to do it as Britishly as possible. The government therefore set up the Independent Television Authority, an organisation that would oversee and extremely heavily regulate

---

* Mostly shouting 'Reeeeeehhhhrrr!!'
† *Do* laugh at the back. This is what they called it and it is funny.
‡ These are only mildly exaggerated versions of the sort of product placement that Americans were used to.

commercial television, with all would-be broadcasters having to abide by strict rules.

1. Adverts and programmes must be separate, and never acknowledge each other.
2. No more than six minutes of adverts per hour.
3. Programmes must be 'British in character' (except foreign imports).
4. TV stations must make boring public service content such as news, local programmes and religious programmes.
5. No adverts allowed during any programmes that feature the royal family.

As well as laying down rules, the ITA had the more fun job of choosing who was allowed to make programmes. The system had to make sure there was competition between programme makers.

Several options were considered for how to make this work. One plan was to have a different TV station operating on each day of the week. Another was to have one company provide pro- grammes for the afternoons, another for the evenings, and another for women.* But eventually the ITA decided that the most sensible idea was to divide the country geographically. The UK would be split into regions, each served by its own dedicated independent television station.

And so, on 22 September 1955, London became the first region to get one, and it was called Associated-Rediffusion. This stuffy, meaningless and appallingly uncommercial name was chosen on purpose. In an unnecessarily over-the-top effort to prove how classy and un-American British commercial television could be, the first programme broadcast on this new station was an hour- long live stream of a seven-course banquet in formal dress at the City of London's Guildhall, with speeches by the Lord Mayor and

* Not a joke.

the Postmaster General. We really do recommend seeking it out on YouTube. It's interminable.

But Londoners needn't have panicked. Once the new company had heavy-handedly proven its point, it got on with the job of doing what it had been set up to do, and made programmes Londoners wanted to watch, with variety shows like *Live at the Palladium*, current affairs shows like *This Week*, and game shows like *Take Your Pick* and *Double Your Money*.

No footage of these early shows survives. Video tape was expensive, and nearly all shows back then were live, so we can only guess what they were like. It's safe to assume they were almost definitely the decadent height of innovative quality entertainment from which it's been nothing but downhill ever since.

The first commercial station had been a success, so it was time for the ITA to roll out more independent TV stations in more regions across the UK.

The next part of Britain to get a TV station was the Midlands in February 1956, when Associated TeleVision (another dull, meaningless name) began broadcasting from Birmingham. Shortly afterwards in May, the north of England got a TV station called Granada (also meaningless, but way more catchy), broadcasting from Manchester.

So far, so good, as these large and somewhat obvious regions hadn't caused too much of a headache. But the ITA had a whole rest of the country to spread television to, and they were about to find out that dividing Britain up sensibly was going to get a whole lot harder.

The ITA's job was to bash fairness into the system wherever possible. So, when it came to carving up the regions, each one would ideally be about the same size, giving the companies and advertisers vying for viewers a level playing field in which to compete. If you were doing that job today you could carefully slice the country up by county or postcode, and make sure every area of the UK got the appropriate local news, served from a headquarters

1. Associated-Rediffusion (London), 2. Associated TeleVision (Midlands), 3. Granada (north), 1956. The first three independent TV stations (they were different at weekends, but that's not important right now).

that represented them best. But unfortunately for the ITA, TV transmitters in the 1950s didn't work like that.

In the olden days, the signal from a TV station was sent down a very long cable to a TV transmitter, an exceptionally tall metal stick that beamed invisible radio waves out into the sky to be picked up by any TV aerial on a rooftop close enough to receive them.* Many of these were hugely impressive structures, taller than the Eiffel Tower, but sadly ignored and excluded from lists of tall things because they don't count as 'buildings'.

Exactly how wide their range was depended on a few things, like how tall the mast was, how high the engineers had turned up the power knob, and crucially, whether there were any hills nearby. Hills are the enemy of TV signals, and they caused no end of problems for the ITA.

In a flat landscape like Lincolnshire, the Belmont transmitter could beam a signal as far as 100 miles away on a clear day,† or 75 miles away on a rainy day, which in Britain is much more relevant. But in lumpier parts of the country, like the north of Scotland, the Mounteagle transmitter near Inverness could only reach as far as, well, Inverness (and a thin sliver of villages across the coast). The more mountains there were in a region, the more transmitters were needed.

Transmitters, and the infrastructure needed to connect them to a TV station (mostly very, very long cables), were expensive. So, the ITA were keen to cover as much of the country as possible with as few masts as possible. This meant they were going to have to make some difficult decisions and awkward compromises that undermined British geography.

---

* Fun tip: if you ever get lost, look up at the rooftops and see which way all the TV aerials are pointing. If you know where your nearest TV mast is, you can work out which direction you're facing. For example, in almost all of London, all TV aerials point south towards the Croydon mast. Sort of unrelated, but it's even easier if you look at satellite dishes, because every dish in the whole of the UK points south. Anyway, that thing about TV aerials was excellent advice in the 1980s.

† These distances are only available in miles because kilometres didn't exist in the 1950s.

TAKING THE 'EAST' OUT OF ANGLIA

The first very mild example of a bad geographical decision (they gradually get worse) took place in, and slightly to the left of, East Anglia.

Next to the Midlands region and up a bit from the London region was a great big patch of the country that needed filling in all the way to the eastern coast. This large patch of England comprising lots of different counties doesn't have a name, or any kind of identity, but it fell to a new TV station to unite them all under one brand. The new station went with the name 'Anglia Television'. This was, geographically speaking, cheeky.

Before Anglia Television began broadcasting in 1957, the word 'Anglia' was almost never seen on its own. The proper name for the round, bumpy bit of England's coast was, and still is, 'East Anglia'. But by naming their station 'Anglia', it made 'East Anglia' sound like it was a subregion of 'Anglia', and that 'West Anglia' must therefore be a thing. But it wasn't, and it wasn't. In reality, places like Northampton and Kettering to the west had nothing whatsoever to do with their unlikely bedfellows in Norwich and Ipswich to the east. It's like starting an ultimate frisbee team in Australia, recruiting players from both New South Wales *and* Queensland, and calling them the New Wales Dolphins. Strewth! And speaking of Wales . . .

WHERE IS WALES?

Having covered lots of England, in 1958 the ITA now turned their attention to Wales. Of all the regions they'd carved up so far, Wales – a country with its own language – was arguably the most deserving of a dedicated TV service. So, starting in South Wales, they built a transmitter at St Hilary in the Vale of Glamorgan. This was a perfect spot to serve millions of viewers in Cardiff,

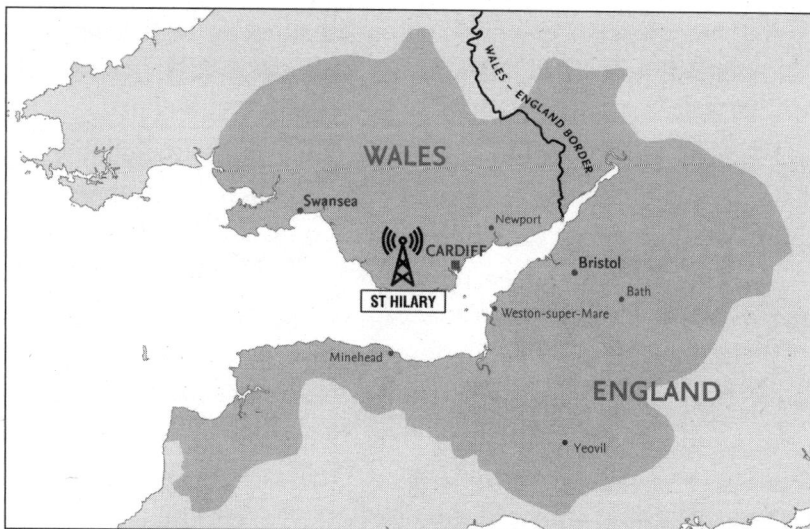

Newport, Swansea and all the densely populated soon-to-be-former mining towns. However, when they switched the transmitter on to test it out, they noticed a problem.

South Wales is very hilly. To make sure all the homes could receive a signal, the mast had to be really tall and really powerful. But South Wales is also bordered to the south by the Bristol Channel, which is a sea and isn't hilly at all.

What this meant was, to broadcast in Wales, they had no option but to send exactly the same broadcast across the sea, crystal clearly, to more than a million viewers in the west of England, including the major city of Bristol.

Rather than build another expensive transmitter on the English side to give Bristolians a very justified television service of their own, the ITA took the more cost-effective approach of declaring that this new franchise area would be for 'Wales *and* the West of England'.

Culturally speaking, the union between these two places across the sea from each other was a very weird and illogical one. The Severn Bridge connecting the two sides wouldn't be built for another eight years, meaning the union was even weirder and less logical back then than it would have been today.

And so, when the new TV station launched on 14 January 1958 (with the necessarily explainy name 'Television Wales and West', or TWW*), viewers in both countries were perplexed.

Folks in Bristol would sometimes tune in to their new TV station, only to see the news in Welsh. Welsh-language programming could creep up on TWW at any time, taking up slots that could have been given to programmes in English – effectively giving Bristol less telly than the rest of England.

Viewers in Wales, similarly uninterested in news headlines about tigers escaping from Bristol Zoo, weren't happy about the compromise either. The creation of this cross-Channel channel diluting their identity meant that the Welsh had been denied a TV station they could truly call their own.

Undermining national borders within the UK would become something of a habit for the ITA.

SCOTT-*ISH*

By 1961 independent TV stations had spread to the far north of the UK, with Tyne Tees Television in Newcastle and Scottish Television (STV) in Glasgow. But this left an awkward gap between the two, with neither station reaching the England–Scotland border.

This gap was so large and so hilly that two enormous masts would be needed to cover it all. The good news was that the ITA achieved this in 1961, with new transmitters built in Caldbeck and Selkirk. The bad news? Neither STV nor Tyne Tees could use them. The problem was that *both* masts covered *both* countries. This left the ITA with no choice but to create a new franchise for a

---

* In case you've glanced at Wales on the map at the beginning of this chapter and are confused about what 'HTV' is supposed to mean, don't worry, the names kept changing all the time, and all is about to be explained in excruciating detail.

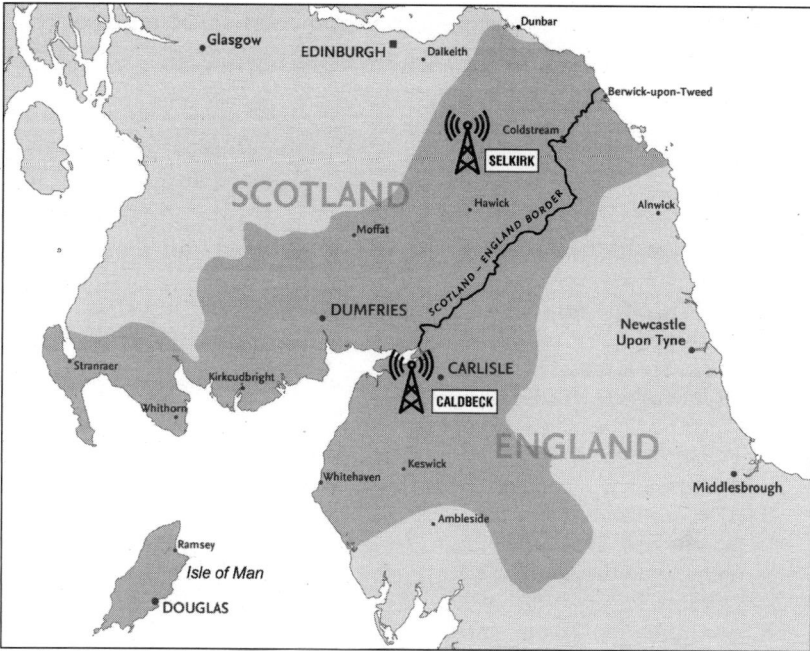

fictional and completely illogical region that straddled both sides of the border, serving Berwick-upon-Tweed (England), Dumfries and Galloway (Scotland), most but not all of today's Cumbria (England) and the Isle of Man (neither Scotland nor England).

Again, a TV station would be created that had to make news bulletins and programmes to cater to all these disparate entities at the same time. The company that won the franchise and began broadcasting in September 1961 called their station *Border Television*, as if admitting that it had no right to exist. Bafflingly unintuitive though this region was, it constituted the final piece of the puzzle, meaning that the ITA had now successfully supplied a TV station to every corner of the UK.[*]

---

[*] Well, technically, the final piece of the ITA's puzzle came one year later in 1962, when the Channel Islands – a collection of Crown Dependencies off the coast of France – got a tiny franchise all of their own with the somewhat tautological (and what would prove decades later to be frustratingly ungooglable) name 'Channel Television'.

After seven years of troubleshooting with transmitters, the ITA's first official map of an entirely televisioned UK now looked like THIS:

4.30 p.m.

1. **WESTWARD:** *Sail Away* – A very gradual zoom out from a large ship.
2. **SOUTHERN:** *Class Act* – Reality show in which a working-class northerner must pass themselves off as an aristocrat. He can't, and order is restored.
3. **ASSOCIATED-REDIFFUSION:** *Knees Up Mother Brown* – Pearly King and Queen beauty contest.
4. **TWW:** *Y Clifton Crogi Bridge* – Documentary about the Clifton Suspension Bridge, but every other word is in Welsh.
5. **ATV:** *Tap Dancin' Time* – Station boss Lew Grade tap dancing on the spot for 45 minutes.
6. **ANGLIA:** *Knightmare* – Kids' adventure show. CG not invented yet, so they all stand in an empty room.
7. **GRANADA:** *Coronation Street* – Soap opera targeted squarely at northerners.
8. **BORDER:** *Eloped* – Reality show following young couples running away to Gretna Green.
9. **TYNE TEES:** *Billy Elliot's Grandpa* – Film about an ordinary coal miner with no unusual hobbies.
10. **ULSTER:** *Giant's Causeplay* – Hexagon-based game show that predates *Blockbusters*.
11. **SCOTTISH:** *Black Pudding* – Studio sitcom about a kleptomaniac Glaswegian shopkeeper starring a young Sean Connery.
12. **GRAMPIAN:** *Ma Wee Nessie* – Documentary about the Loch Ness Monster's mum.
13. **CHANNEL:** Who cares? Only three people are watching.

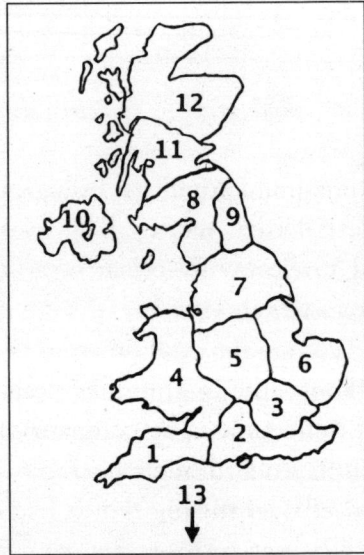

The ITA's 13 regions in 1962, with what each one might typically be showing on a Thursday at 4.30 p.m.

THE MAD MAP STARTS TO MATTER

The more households bought tellies, the more important the ITA's map became. In the days when the TV was the only source of entertainment in the home, and there were only two channels* showing programmes that the whole family, or classroom, or offshore oil rig watched together, your local independent TV (or 'ITV') station played an enormous role in understanding your regional identity. For some, the TV station became synonymous with the region it served. 'Granada', whose name had been chosen almost at random because it sounded Spanish and exotic, was now a word that the whole country associated with Manchester and the north of England. Travelling to another part of the UK and seeing an unfamiliar ident† on your nan's TV was an exciting reminder that you'd strayed from home territory. The local news told you which places were yours to care about. Indeed, the regions were sometimes pitted against each other. Contestants on *Bullseye*, a darts-based game show, were introduced from their respective ITV region. The *Disco Dancin' Championship* had a dizzying title sequence in which sequin-clad hopefuls thrashed about on a glittery stage for ten seconds each next to the logo of their local ITV station. The bi-annual 'Telethon' (ITV's short-lived and very forgotten answer to the BBC's much more successful Comic Relief) displayed a running total of how much money for charity had been raised over the phone by viewers in each region.

The nation soon took this map to heart, despite it being based on nothing historical or cultural that had come before and was guided by nothing more meaningful than a few hills.

But the map was about to change.

---

* Three after 1964, when the BBC launched a second channel, imaginatively called BBC2.

† Animated logo with an accompanying jingle that reminded you what channel you were watching, representing the region as well as the then-prevailing trends in graphics, style and music. An irresistible YouTube rabbit hole.

THE MAP GETS MEDDLED WITH . . . A BIT

1967 was the year of reckoning for the first generation of ITV companies – when their franchises were due to expire. Most of them were deemed good enough to get permission to keep broadcasting for another ten years (apart from TWW, who'd annoyed both sides of their patch by being based neither in Bristol *nor* Wales – their head office was in London). They were replaced by Harlech Television (HTV), so named because the head of the company was Conservative politician Lord Harlech. They really didn't give a monkey's about good names back then.

This 'franchise round' gave the ITA their first opportunity to meddle with the map they'd made and see how good a job they'd done of dividing the country fairly. When they stared at the map, however, they realised they'd completely misunderstood their brief.

The regions looked about the same size from space, but their populations varied enormously. Border's patch, for instance, contained barely a million viewers, versus Granada's more than ten million. If you lived in Blackpool, you were ten times less likely to appear in the background of a local news bulletin than someone who lived in Berwick. The ITA concluded that two of the regions, London (Rediffusion's patch) and the north of England (Granada's patch), were too big.

The north was easy to solve. They simply split the region in two, letting Granada keep the western half, and gave the eastern half to a new company based in Leeds calling itself Yorkshire Television. This was a good idea, for now, but would prove to cause a big problem later. Stay tuned . . .

London, on the other hand, was a much harder nut to crack. As the country's only mega-city, its ten million people crammed into an area covered by a single mast, there was no way to avoid it being the most populous by a country mile, giving whoever controlled the Croydon transmitter a hugely unfair advantage.

Unable to split the region geographically, the ITA went with a slightly mad approach.

SOMETHING FOR THE WEEKEND?

From 1968 the London region was to be the only one in the country shared by two TV stations operating on the *same* channel.* From Monday to Friday, London's ITV station would be Thames Television. Every Friday at the arbitrary and ludicrously specific time of 5.15 p.m., Thames Television would say goodbye and when the commercial break was over, an ident for a completely different station would appear: London Weekend Television.

This surreal switcheroo had its pros and cons. On the plus side, Londoners uniquely got to enjoy the sensation that the TV set was taking its school uniform off – the bold, stripy letters 'LWT' painting themselves across the screen providing a clear signal that the weekend had arrived. On the minus side, however, it was harder to plan what you were going to watch. The two companies, competing for the same advertisers, were sworn enemies, refusing to promote each other's shows or even acknowledge each other's existence. The Thames Television announcer (in those days, announcers were in vision, sat behind a desk and wearing a tie) would sign off with a deliberately misleading message – 'That's it from Thames this week, we'll be back on Monday'– in a bid to make viewers think it was time to switch their set off and go to bed. It affected scheduling for the whole network too. The country's best-loved comedy double-act, Morecambe and Wise, were unable to do their regular hugely popular Christmas show in 1982 because Christmas Day happened to fall on a Saturday.

---

* In the very early experimental days of ITV, London, the Midlands and the north were all split into weekday/weekend franchises like this, but London was the only one where this practice was maintained. We have to clarify this, or the massive telly nerds will spot it and send us angry letters.

But a huge opportunity for change, proper fairness and a map that made sense would come along for the ITA at the end of the 1960s, brought about by a leap in technology. And, in a predictably British way, this opportunity was spectacularly missed.

RED AND YELLOW AND PINK AND GREEN

Despite the real world changing from black and white to colour about a third of the way through *The Wizard of Oz* in 1939, this exciting change – making programmes more immersive and giving snooker commentators less to talk about – wouldn't hit British televisions until July 1967.

Colour TV used a completely new ultra-high frequency broadcasting system. We won't go into the physics (partly because you don't need to understand it, and mostly because we don't), but the important thing was that the ITA now needed to build a whole new generation of TV masts – and there had to be three times as many of them as before.

At last they had an opportunity to completely redraw their bonkers map. No longer constrained by the broad uncontrollable footprints of a small handful of massive old-fashioned transmitters, they could fine-tune the boundaries and slice the country fairly. Except . . .

Colour TV technology was expensive. Not everyone could afford it all at once, and that went for the TV companies as well as the viewers at home. Colour on ITV had to be rolled out gradually across Britain, and the two systems would have to co-exist for decades.*

And so, to facilitate the smooth transition, the ITA had no option but to try their best to place the new generation of transmitters to

---

* It would take until 1985 for the last black and white transmitter to be switched off. For one of us, that's living memory. For the other, it's ancient history.

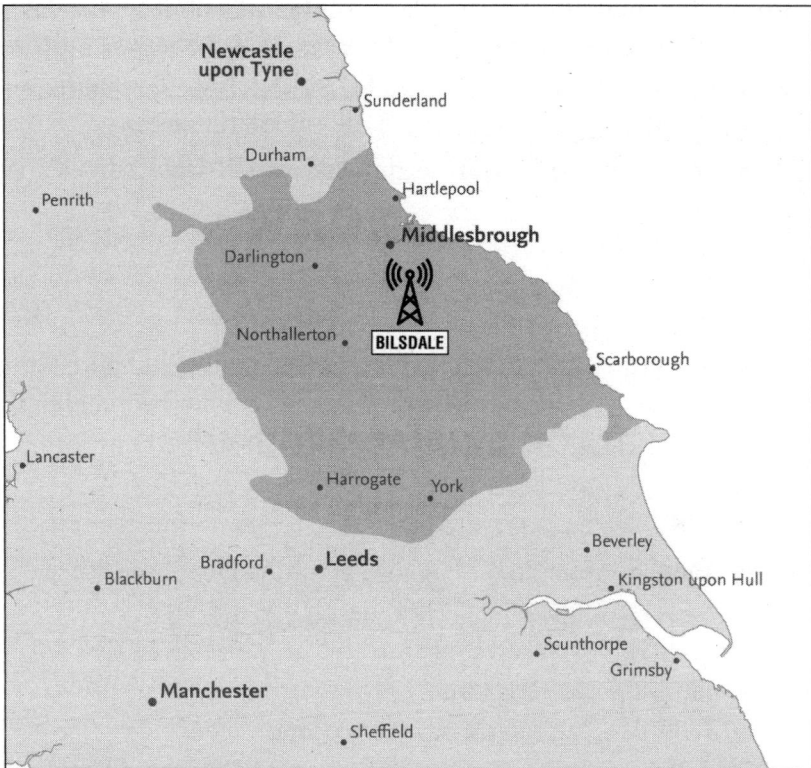

cover the same daft, illogical areas as the old ones. Northampton was still in 'the East', 'Border Television' was still a thing, and the West of England was still in Wales. It was bad enough that the Welsh had had to share with the English, but now they had to share with the English for no good reason.*

Worse still, unable to absolutely perfectly match the original illogical compromise of a template they were working with, the new system introduced yet more anomalies and awkward compromises. And nowhere did this cause more irks than the north-east . . .

---

* To be fair, by this time they'd managed to separate this dual region into separate areas for the news – known as 'Wales' in the west, and 'West' in the east, but as far as game shows like *Bullseye* were concerned, they were still one big, homogenous semi-Welsh blob.

In 1971, a new UHF colour transmitter was built at Bilsdale, near Northallerton. While it was able to reach more than a million homes, it also happened to be right on the boundary between Yorkshire and Tyne Tees's turf.

This predictably led to an argument, which went something like this:

Yorkshire: (gruff Yorkshire accent) We should get t' transmitter.

Tyne Tees: (melodic Geordie accent, in which it's really satisfying to say 'transmitter')
     No way, man, we should get the transmitter.

Yorkshire: It's int' middle o' North York Moors. It's got Yorkshire int' bloody address.

Tyne Tees : So what, man? It's, like, ten miles away from the River Tees. You cannae spell Tyne Tees wi'out Tees.

Yorkshire: This mast serves bloody York, and people in York should get proper Yorkshire Telly.

Tyne Tees: Aye, but there's way more people on our side. Middlesbrough, Hartlepool, Darlington, all the way up to Sunderland on a sunny day.

Yorkshire: You've never had a sunny day.

IBA: Have you considered asking the locals which one they'd prefer?

Yorkshire and Tyne Tees: Who are you?

IBA: Oh yes, sorry. We used to be the ITA, but now we're in charge of commercial radio too, so we've swapped the T for 'telly' to B for 'broadcasting'.

Tyne Tees: They won't let us use our mast.

Yorkshire: No, they won't let us use *our* mast.

IBA: Gentlemen, settle down. As far as all of us down in London are concerned, you're basically the same thing anyway. I'm sure we can find a sensible way to sort this out.

So the IBA came up with a compromise. A bad one.

The Bilsdale transmitter would be given to Tyne Tees,* but Yorkshire Television were handed control of the Belmont transmitter down in Lincolnshire by way of compensation. This transmitter was taken away from Anglia TV – who weren't involved in the argument and hadn't done anything wrong – and it ultimately resulted in this map's most horrendous sin.

Due to the flatness of the eastern side of the country, the boundaries of the new Super 'Yorkshire' had now shifted all the way down to Norfolk, meaning King's Lynn, a posh, genteel and decidedly southern town, was now receiving the same local news as Bradford.

Ignoring the England–Wales border? Cheeky. Ignoring the England–Scotland border? Pretty audacious. But breaching that holiest of boundaries, England's north–south divide? Positively unforgivable. The no-longer-accurate name 'Yorkshire Television' emblazoned on southern screens only made it much worse.

Chilled at the prospect of their children hearing the words 'ee by gum', some 70,000 East Anglians complained to the IBA. The issue was raised in parliament, and the *Guardian* reported that some people were going to 'considerable trouble and expense to continue receiving Anglia Television'. Those who were able to do so climbed dangerously onto their rooftops to extend and rotate their aerials, while those who weren't squinted defiantly at fuzzy pictures of what they believed should have been their local news programming.

'Daaaad? Why are we watching a snowstorm?'

(*thick East Anglian accent†*) 'Because it's a good, honest East Anglian snowstorm! Now be quiet and eat your Bernard Matthews turkey breast with hot English mustard!'

---

* Tyne Tees were in more financial trouble than Yorkshire, so the IBA wanted to give them a leg up. So goes the theory, anyway. The IBA didn't need to explain their actions to anybody.

† Your guess is as good as ours.

'It hurts my tongue.'

'Northern softie!'

Yorkshire Television did try to win over their new southern audience by changing their idents to include a map of their new catchment area, which if anything served only to highlight how sprawling and absurd the area was.

But pretty soon, none of it was going to matter anyway.

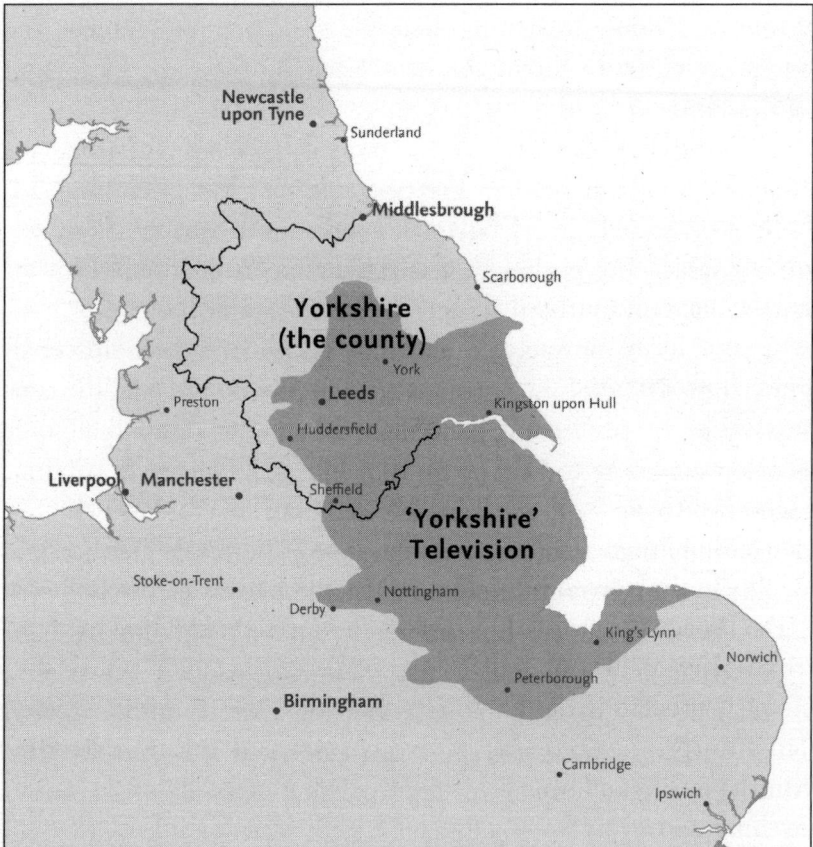

Yorkshire Television's broadcast area from 1974. Rather a lot of this is not Yorkshire.

THE REGIONS START TO DISAPPEAR

For the first three decades of its life 'independent television' hadn't been a channel, it had been a concept, a broad category containing many TV stations with only one thing in common – they weren't the BBC.

But when competition came along in 1982 from a new fourth channel – which, in a burst of unspiration, was called Channel 4 – as well as a slew of satellite and cable channels throughout the same decade, the ITV network came under pressure to band itself together and save costs.

ITV now began a slow and gradual process in which more programmes were networked nationally and fewer programmes were shown in just their home region. What British viewers ended up with by the late 80s was an ITV schedule that was pretty much exactly the same up and down the country, the only variations being the local news, local adverts and the idents that came between programmes. It was only a matter of time before ITV's heritage as a regional network would disappear, and it would simply turn into just another TV channel.

The lengthy drama-filled process of how the network ate itself and transmogrified into the channel it is today is a long, complicated one. So, in the interest of brevity, we'll keep it to one sentence.

The Broadcasting Act 1990, which had been introduced by then Prime Minister Margaret Thatcher to deregulate ITV, which she famously proclaimed to be 'the last bastion of restrictive practices' – an unsurprising move for a politician whose dogmatic obsession with the purity and sanctity of the free market led her to privatise anything and everything she could get her hands off – saw two major changes to how the network was run, the first of which being how the IBA, by now renamed to the ITC, chose which companies were allowed to broadcast, which until then had been run like a beauty contest, where on a regular-ish basis of usually every ten years (some exceptions being made when the network had more tricky things to worry about, such as introducing colour television technology or excessive industrial action, something Thatcher was particularly annoyed about) the different companies would pitch their case to the ITC based on their ability to broadcast, their suitability for their respective region, their ideas for the future, and, if they were able, their past record, and the ITC would pick the company who'd made the case that they were the most capable of serving both the viewers in their respective region and the network at large, but under the new system the winner was instead chosen by way of a blind auction whereby various potential broadcasting companies made their case to the ITC based entirely on how much they were willing to pay the Treasury on an annual basis, leading to an infamously chaotic franchise round in 1991 in which no company had the foggiest idea how much their neighbours or potential rivals were bidding, with some as high as £45 million and some as low as £2,000 (because they'd done some sleuthing and found out they were bidding unopposed), which resulted in several surprising outcomes that sent shockwaves through the broadcasting industry, including TSW losing the south-west, TVS losing the south, and most famously of all, Thames Television, the largest and most reputable broadcaster on the network, responsible for such hugely popular shows as *This Is Your Life*, *The Bill* and *Minder*, as well as much beloved children's programmes such as *Rainbow* and *Count Duckula*, losing their London weekday franchise to a hitherto unknown entity called Carlton Communications – one of whose directors was future prime minister David Cameron – who had managed to outbid Thames with an overly large bid of £43.2 million, which many suspected to be an intended outcome by the Thatcher government, which had very publicly fallen out with Thames Television after the 1988 broadcast of an episode of current affairs show *This Week* entitled 'Death on the Rock' – a groundbreaking, controversial and ultimately award-winning documentary that called into question the actions of the Thatcher government when ordering the shooting of three members of the IRA in the British Overseas

Territory of Gibraltar, and the result of all this was a network that from January 1993 onwards comprised an unhealthy mix of old broadcasters who'd been rendered unable to keep up the quality of programmes they'd built their reputations on as a result of the punitive and arbitrary dues they now had to pay to the Treasury, and brand new companies with no prior experience in broadcasting who had no option but to source their shows from smaller independent production companies, beginning ITV's inevitable journey towards cost-cutting programming, and the second major change to ITV brought about as a result of the Broadcasting Act 1990 was that companies were, for the first time, permitted to own a stake in more than one ITV company, which had the inevitable and very much intended consequence that ITV companies were now able to buy each other out, resulting in a takeover frenzy resembling a nationwide game of *Regional TV Monopoly* that lasted throughout the 1990s, where Scottish TV merged with Grampian, Yorkshire merged with Tyne Tees, Carlton bought Central and rebranded it to Carlton Central, then it bought Westcountry and rebranded it to Carlton before buying half of HTV, while at the same time Granada bought the other half of HTV, LWT, Meridian and the newly merged Yorkshire Tyne Tees before buying Anglia and Border, resulting in just two giant companies, Carlton and Granada, running all the regions in England, which led to the unavoidably inevitable outcome of the two companies merging in 2004 to form ITV plc, which proceeded to remove all trace of the regional identities, brands, programmes and logos, replacing them with a generic nationwide identity, using the name 'ITV' on its own on screen for the first time, limping into the new millennium, its relevance diminishing with every passing year, and making only the bare minimum of regional content, which is almost exclusively news, and developing a reputation for making hard-hitting discussion shows like *Loose Women*, compelling nature documentaries like *I'm a Celebrity Get Me Out of Here*, and *Tipping Point*, which is rubbish.

THE PRESENT DAY

It's hard to say whether ITV's homogenisation, de-regionalisation and ultimate decline was Margaret Thatcher's fault, or an inevitable consequence of the changing media landscape. Blaming Thatcher for any of this country's problems is of course easier, and a national pastime.*

A reassuring consolation for nostalgic fans of regional telly is that, despite the company that owns all the franchises being called 'ITV', and despite the name 'ITV' being used on screen, in a *technical, legal* sense the regional stations – along with all the names that haven't been seen or heard for decades – still exist, but only in the same meaningless theoretical way that the African Association does (see previous chapter).

Today, channel 103 on television sets all over Britain is called 'ITV1'.† Some very minor adjustments have been made to the regions, with Bristol finally detached from Wales after more than fifty years. But it now hardly matters. With the exception of local news bulletins and local adverts, it plays exactly the same schedule up and down the country in every region . . . Except one.

Scottish TV, despite all the takeovers that took place south of the border, resisted and held out against the invaders. To this day, in Scotland, the channel that the rest of the country calls ITV is called STV, which plays a mix of nationally networked programmes and shows just for Scotland.

However, due to a mishap dating back to trouble with transmitters back in 1961, the Scottish Lowlands, even now, do not receive STV, getting English ITV instead, which is at best weird and at worst extremely unfair when there's Scottish football on, or a

---

* Mark spent 15 minutes looking for his keys this morning, and was muttering, 'Bloody Thatcher, where's she put them now?!'
† We were astonished to learn during the research for this chapter that there's an ITV7, just for horse racing.

Gaelic soap opera, or a news story about the Scottish parliament. Throughout 2014, when STV News was banging on about the upcoming Scottish independence referendum, these bulletins weren't seen in the Scottish borderlands, where people would arguably be most affected by the outcome.

Sadly, or maybe thankfully, this messy map seems to have become irrelevant before it'll get another chance to be fixed. Regional programming shows no sign of making a comeback, and even national programming is under threat in an ever more glo-balised, streaming-focused landscape, where a broader (arguably blander) international appeal has become a commercial necessity. As more people turn elsewhere for their news, and linear TV con-tinues its steady march to full digitalism, the lines on this map seem much more likely to disappear entirely before they get further meddled with again.

A MAP WE CAN LEARN FROM?

While ITV's regions map was massively flawed and its borders were put, either through choice or necessity, in some pretty bonkers places, there's a lot we can learn from it, and it has a good few advantages over other more traditional ways of dividing up the country.

For one, it includes the Channel Islands – a refreshing change from the norm for inhabitants of Jersey, Guernsey, etc, who have to read the words 'excludes Channel Islands' several times a day. The Midlands, the splodge of the country that neither northern-ers nor southerners want to claim as their own, is treated as one region and not split somewhat arbitrarily into East and West, as is needlessly, divisively done in every other map. ITV's 'London' is far broader than the 'Greater London' area, that small octago-nal-ish shape shown on most other maps, which has, since 1965, defined what politically counts as London. It provides a much more accurate reflection of the vast pull of the capital, influencing

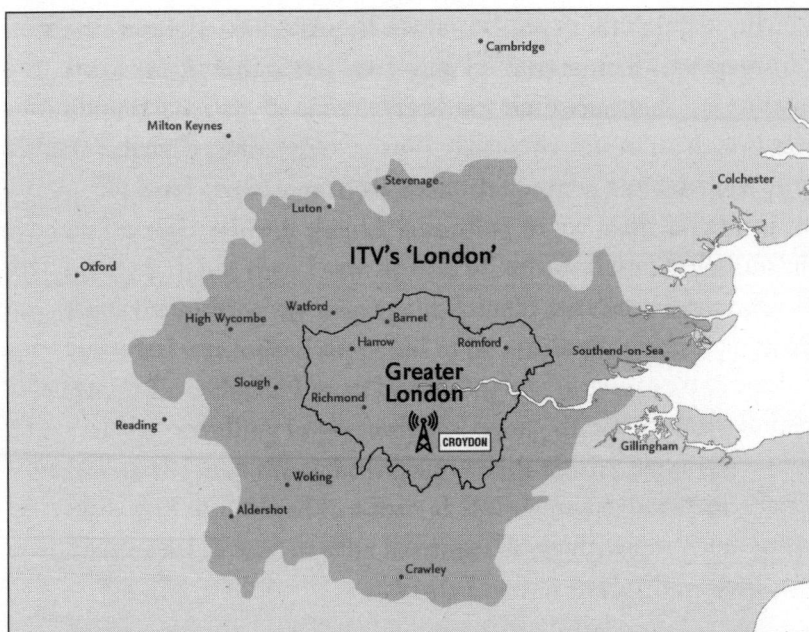

everything from commuting patterns to job opportunities to house prices.*

But perhaps most importantly of all, the boundaries are blurry. While the ITV regions map purports to have clear lines dividing the regions, in reality there was a tremendous amount of overlap between the transmitters, and a great number of households were able to tune in to more than one ITV station. The fact that fuzzy areas existed in which you could rotate your aerial and allegiance whichever way you fancied, as well as countless pockets of tiny anomalies across the landscape, is a truer reflection of how a nation culturally divides itself than unrealistically clear, clean lines, whether it's what local TV you watch, what football team you support, what time you eat your dinner, whether you even call it 'dinner', or how you pronounce 'lucky bath'.

* If we all went by ITV's definition of London it would end the silly arguments about whether Luton, Gatwick and Stansted count as London airports. (They do.)

Maybe this more porous approach to borders is something that future governments could learn from. Administrations arguing over where their jurisdiction extends could do worse than going by the broadcast reach of a tall transmitter, which arguably makes only a little less sense than the arbitrary, artificial and equally absurd solid lines we've been drawing across uninhabited deserts on traditional maps for only the last few hundred years.

Or, at the very least, it might be fun to have a different government at the weekends.

6

IS A GLOBE A MAP?

No, it's a globe.

7

THE FICTIONAL MAP
THAT BECAME REAL, THEN
FICTIONAL AGAIN (AND
THEN REAL AGAIN
(BUT ONLY FOR A BIT))

Good artists copy; great artists steal.
Pablo Picasso. Followed by Andy Warhol,
Tony Hart and Neil Buchanan

Mapmaking, if you do it properly, is a time-consuming and money-consuming business. You have to do field surveys and travel to multiple locations, some of which can't be accessed in a vehicle. You have to measure distances, angles and elevations using heavy, cumbersome equipment like theodolites, levelling instruments and very long tape measures. And once you've gathered this data, it all has to be painstakingly compiled using either computers or triangulation and maths, depending what century it is.

This is a complicated enough process when just drawing a map of the basic shapes of an area, but the job gets even harder if your map needs to include labels. Even if you've got your own satellite photography kit knocking around, which – by the way – is even more expensive to maintain than tape measures, this will tell you nothing about the names of all the towns, streets, parks, schools and so on. Getting information like this requires yet more careful surveying. And what of all the important features that your map needs to show that are invisible from the ground, like administrative boundaries, or roads that have been planned but not built yet? Obtaining such data requires liaising with various government departments, some of whom don't respond to emails for weeks.

So, with all this in mind, it's easy to see why the temptation is so strong to *not* do mapmaking properly. Instead of going through all that rigamarole, why not simply trace the information you need from a map that's already been made? After all, how would anybody know? Two separate companies surveying the same real world should end up with exactly the same results.

Plagiarism is a genuine problem for mapmakers, one that only gets worse the more detailed their map is. The closer it reflects the real world, the harder it is to prove that there has been any creative process at all. So, when cartographers have gone through all the extensive and expensive effort to gather the data for their maps, how do they protect themselves from other companies copying it for themselves and passing it off as their own, with slightly thicker lines, bolder colours and a chunkier font?

It turns out, mapmakers do have a trick up their sleeve to stop this from happening. It's a centuries-old method that enables them not only to spot when their work has been stolen, but also, if necessary, to be able to prove it in a court of law.

The idea is . . . and it's one of those ideas that's so stupid it disappears off the stupid side of the stupidity graph and reappears on the clever side . . . you make your map wrong on purpose. All you have to do is put a deliberate mistake on your map that definitely isn't in the real world. If that same mistake turns up on someone else's map, you'll know the only place they could have got it from is your map, and . . . busted!

Incidentally, this concept where creators add subtle little incorrect details to protect their copyright isn't just limited to maps. You can (or, if they're doing it right, you *can't*) find made-up words in dictionaries, fictional entries in encyclopaedias, fake phone numbers in phone books, non-existent businesses in business directories, meaningless strings in software code, extra screws in architectural plans, bad advice in medical textbooks and glaring factual errors in light-hearted books about maps.

That's right. Consider this a warning to any would-be authors out there. One fact in this chapter isn't true.* Not saying which one, obviously, but if we spot it in your book we'll know what you've done, and you'll be getting a tersely worded email from our friend's mum, who's a lawyer.

Copyright traps on maps can take many different forms. Of course, they don't work if they're massive howlers like putting New York on the west coast of Africa, or turning Japan inside-out, or adding a dense network of protected cycle lanes across the Royal Borough of Kensington and Chelsea. The trick is to add a false detail that's small and subtle enough to go completely unnoticed and cause no trouble to the map user, but easy enough to identify if it turns up on somebody else's map. Getting this balance right is a real skill. It also sounds like a right hoot.

Sometimes it can be a harmless spelling mistake. 'Book Mews' in central London is shown in the iconic *London A–Z* as the more plausible sounding 'Brook Mews'. But it's a tiny dark alleyway off a side street that only three people live on, so it doesn't cause any nuisance. It's rumoured that the *A–Z* has at least one 'mistake' like this on every single page.

Sometimes map traps can be subtle alterations in the physical make-up of the map, such as bending a mountain contour the wrong way, or making a very squiggly road slightly squigglier. A 2011 map of the Swiss Alps produced for Swisstopo by cartographer Paul Ehrlich has contorted contours that look like a marmot climbing up the side of a mountain. This was drawn just before his retirement, so it was probably done for the purposes of mischief rather than copyright protection. What were they gonna do, fire him?

But by far the most well-known, easiest to prove and most fun type of deliberate mistake is a feature such as a building, a street or even an entire town that simply doesn't exist. Towns such as this that appear only on paper are known as 'paper towns'. And they are everywhere.

* We can only apologise if there turns out to be more than one.

An example of a map with a subtle mistake. Can you spot it?

The 1978 edition of the official state of Michigan map shows the fictional cities of 'Goblu' and 'Beatosu' in the thin strip of neighbouring Ohio at the bottom of the page – the names being a not at all subtle dig at the University of Michigan's rival Ohio State University. ('Go Blue' and 'Beat OSU'. Get it?)

The town of 'Argleton' in Lancashire existed only on Google Maps until 2010, when it was quietly deleted from their database, probably because word started to spread when somebody clicked the 'satellite view' button and revealed it for the empty field that it was. (They might also have spotted that 'Argleton' is a somewhat unsatisfying anagram of 'Not real G'.)

And those are just a handful that we *do* know about. Paper towns are like background extras in movies – if you notice they're there, they're not doing their job properly.* That means there are countless more out there, but, by their very nature, we don't and can't know about them all. Many of them are hidden in old maps whose cartographers are long dead, and so they may never be discovered.

Paper towns are also a bit like car alarms or nuclear warheads – while it's nice to know they'll do their job in the worst-case scenario, their main reason for existing is to act as a deterrent. Actual court cases where plagiaristic defendants stand in the dock sputtering and blustering over blatant copying are disappointingly rare. Mapmakers, knowing that there are hidden traps in their rivals' work, tend to stay away from copying each other. All the different companies make their own maps from scratch, sending their surveyors to the same locations in an inefficient atmosphere of mutually assured litigiousness.

However, among all these paper towns there's one that stands out from all the rest, one that failed so miserably in its mission to go undetected that it became the best-known paper town in map history, with the story accompanying it becoming a firm favourite among trivia nerds, map geeks and funny-fact fans. In fact, such is its status in the canon of map folklore that various sources bicker about

---

* English teachers, feel free to use this sentence in a spelling test.

when things happened, who they happened to and whether they even happened in the first place. One thing we do know for certain is that for this story we must travel across the Atlantic to New York.*

THE BIG THREE

For most of the 20th century three companies were responsible for the vast majority of all road maps produced in the United States. The 'Big Three', as they were double-accurately known, were, in decreasing order of bigness, Rand McNally, H.M. Gousha and General Drafting. General Drafting, the least memorably named of the three, was run by Otto G. Lindberg, a force to be reckoned with in the mapping world. Originally from Finland, he arrived in the United States in 1909 and, with a loan of $500 from his father, set up his own business on Broadway in the heart of New York City. Although many of his customers came in expecting a musical, they were pleasantly distracted by his excellent maps, which quickly gained a reputation for clarity, accuracy and beauty, even if they did lack the sort of emotional catharsis some of the clientele had been hoping for.

Unlike his two bigger rivals who sold their maps to lots of smaller companies, Otto's General Drafting had only one customer, but it was a big one. In 1923 he managed to persuade Standard Oil of New Jersey – the company that would later become Esso – to let General Drafting be their exclusive map publisher. Right up until the 1970s, gas stations in America used to give out road maps for free, and with Esso boasting one of the largest chains of gas stations in the country, this made General Drafting's maps among the widest distributed, best known and most trusted in America.

---

* Not necessarily across the Atlantic, depending where you're reading this. If you're already in New York, just wait for us and we'll carry on the story when we get there.

Like most mapmakers, Otto was not averse to hiding copyright traps in his maps. In 1925 he published a map of Delaware County, New York, with his assistant Ernest Alpers, and it included a cheeky little paper town.

The scale and level of detail on the map meant that a paper town was the perfect copyright trap, because all that was required was a dot and a made-up name, no need to actually doodle any fiddly streets. They placed their naughty dot in a quiet corner of the Catskill Mountains – the scenic region where New Yorkers used to spend their summer vacations before flying to more exciting places became affordable – on a dirt track just north of the tiny hamlet of Roscoe, right on the border between Sullivan County and Delaware County.

This was an ideal location for a paper town – a small road, leading off a small road leading off a small road that has since been bypassed by a bigger road. It wasn't on anyone's way to anywhere in particular and therefore very unlikely to cause any problem to passing motorists, if any ever came. Now, all the paper town needed was a name.

Otto G. Lindberg and Ernest Alpers chose to combine their initials, OGL and EA, and jumble them up to form 'Agloe'.

An inspired choice, both vain and subtle at the same time, and an impressive feat to come up with a plausible American town name when most of the letters available were vowels. Would be a great starter for Wordle.

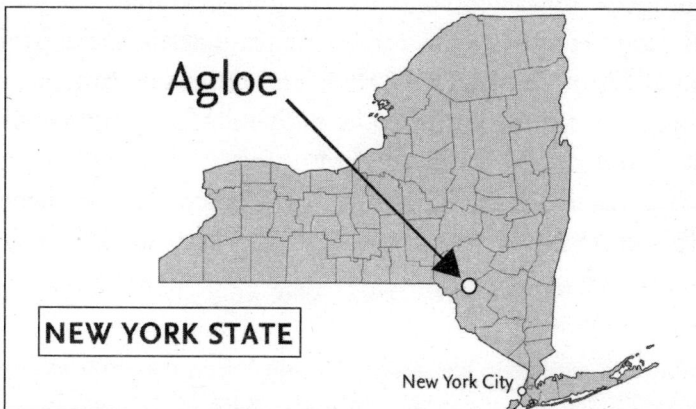

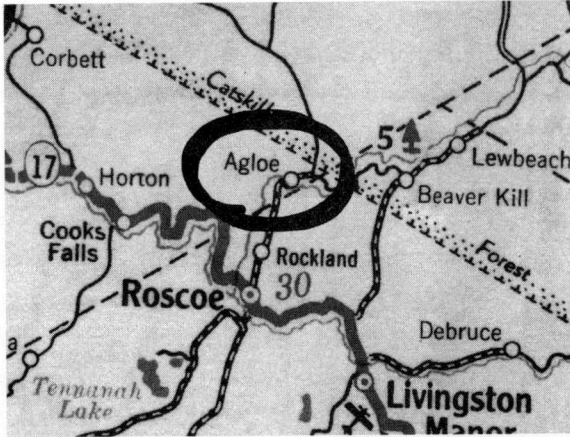

This is General Drafting's 1939 map containing Agloe. We've drawn
a big circle round it so you can see it. *They* didn't do a big circle. That
would have been something of a giveaway.

Precisely what happened next depends on which version you
believe has more verisimilitude. Happily, both lead to the same
narrative result, so the chapter should remain satisfying despite
the factual murkiness. The most popular version takes place in the
1950s and goes like this . . .

Otto G. Lindberg was browsing through his local library in
Manhattan looking for hotdog recipes* when he decided to have a
quick flick through the map section. He knew his own maps were
the best, but it made good business sense to keep an eye on what
the competition was up to, as well as giving him a sense of smug
satisfaction. He picked up a copy of the latest edition of Rand
McNally's map of New York State and thumbed through to the
page he always turned to when he did this. His cigar tumbled out
of his mouth and fell to the carpet.†

There, just north of Roscoe, in the exact same spot where he
and Ernest Alpers had put it three decades earlier, but in a slightly
different font, was 'Agloe'.

---

* We did only say it goes 'like' this.
† Smoking in libraries was mandatory in the 1950s.

OTTO MAKES A PHONE CALL

'Hello, you're through to Rand McNally Publishing, how may I help you today?'

Otto was sitting in his office late at night, Rand McNally's map folded open on one side of the desk, Agloe having been aggressively circled in thick blue cartographer's pen,* and the old-timey telephone on the other, a lit cigar balanced horizontally between his nose and top lip like a moustache. He'd waited a long time for this moment.

'Listen, McNally,' said Otto, 'you're treadin' on thin ice, and it's warmin' up fast.'

'Excuse me?'

'You're knockin' on a door you don't wanna open, buddy. You're settin' the table for trouble, and it's eatin' time.'

'Can I help you, Sir?'

'Quit playin' dumb, mooch. You put your chips on the wrong number, and the wheel's spinnin'. You're fishin' for sharks in a dinghy with fishsticks for paddles, you hear me, bub?'

'Can I ask who it is you want to speak to?'

'It's about Agloe, numbnuts! We found Agloe on that rag o' yours you call a map!'

'OK, just one moment, Sir.'

Forty-five minutes later, Otto was put through to the right person.

'You ripped off our map and we can prove it. There is no Agloe. We made it up out o' thin air. If you got Agloe from our map, which you *must* have done and you know it, that means you got everything else from our map too. We've caught you red-handed, Rand McNally! It's breach of copyright, and we're gonna sue you for every tape measure you got!'

'OK, sorry to hear that, Sir. Would you like my supervisor to look into this for you?'

---

* Normally used for rivers.

Otto hung up and took a big puff on his cigar; he was looking forward to his day in court. After all, this was an open and shut case. Rand McNally would be exposed for the fraudsters they demonstrably and undeniably were, and General Drafting would get compensation. Maybe even a bit of positive press too.

But then, quite unexpectedly, a few days later Otto received a phone call.

Ring ring.*†

'Hello, Mr Lindberg? I'm calling from the offices of Rand McNally Publishing.'

'Weeeeeell, if it ain't the lying, thieving map copycats Rand McNally. Lookin' to settle, are ya? Well, you're barking up the wrong goose and it ain't gonna fly. You jangled with the wrong jump leads and this battery's loaded.'

'Yes, hello again. I've just spoken to our lead cartographer about Agloe, and he says it's a real place.'

'What? No, it's not.'

'Yes, it is. We got the information from the Delaware County records.'

'Huh? That can't be right. Me and Ernie made it up. It's made out of the letters in both our names. We argued about it all night. I wanted to call it Fully Otto-Matic.'

'Sir, I assure you Agloe is a real place. We surveyed it ourselves. You're welcome to drive up there for a visit to check.'

'There is no Agloe! I'll Agloe you in a minute, you meatballs!'

'You can't claim a copyright trap on a map if it's a real place. Goodbye, Mr Lindberg.'

The line went dead, but Otto continued yelling inaccurate New Yorkisms down the telephone long into the night.

The next morning, Otto went on a road trip. Using both his own map and Rand McNally's, he drove out of Manhattan,

---

* Editor: This story takes place in America, so it wouldn't have gone 'ring ring' – it would have gone 'riiiiiiing'.

† Authors: This note alone has made the decision not to self-publish worth it.

crossing the Hudson River over the Tappan Zee Bridge* into the autumnal depths of Upstate New York.

The roads grew quieter, narrower and twistier as the Catskill Mountains began to climb out of the landscape.

After four hours he arrived in Roscoe, continuing to drive slowly, cautiously north, not quite knowing what to expect. The road gently meandered, until a small, lonely building appeared from round a bend. Otto pulled over and got out of the car. His cigar slid backwards down his throat unnoticed as he stared up open-mouthed at the sign on the awning – 'Agloe General Store'. Otto pushed open the door with a ding.

'A customer! Oh my god, an actual customer! Uh, welcome to Agloe, Sir. How may we help you today?'

'You the owner of this joint?'

'Yeah.'

'You come up with the name?'

'You mean . . . "General Store"?'

'I mean Agloe! Why does it say "Agloe"?'

'This is Agloe, Sir.'

'No, it isn't! What's wrong with everybody? There's no such place as Agloe and there never was.'

An angry voice came from behind the stock room door.

'See? I told you there wasn't a town here!'

'Then why was it on the map, honey?'

Otto's face fell. He immediately figured out what must have happened. The shopkeeper, with his trademark poor judgement, continued explaining anyway.

'We'd never heard of this place either. Story is, my wife and I always wanted to open a store in the Catskills. We were driving around and we found this beautiful spot. We looked at where we were on the map, and it said we were in a town called Agloe. So we called it the Agloe General Store.'

---

* This had just been opened and was therefore on Rand McNally's map but not his, which made him even angrier.

The shopkeeper pointed to a crinkled old map that had been inelegantly unfolded and blu-tacked to the wall behind him. In the bottom corner were the words 'General Drafting', 'Esso' and '1925 edition'. And in the dead centre, lovingly circled in thick black shopkeeper's pen,[*] in Otto's favourite font, was 'Agloe'.

Otto scrunched his eyes up and pinched the top of his nose. Thanks to these shop-building bozos, Agloe had now become a real town. Or village. Or – well, point was it wasn't *nothing*. It had a population, albeit a very small one. Still, Agloe was now undeniably a *place*. And crucially, one that robbed General Drafting of any chance of winning a case against Rand McNally.

Suddenly, a tiger that had escaped from the Bronx Zoo leapt in through the window and ate Otto for lunch. And that's how he died.

And that's the famous story that's been retold various times in trivia books, TV panel shows and numerous websites – including Rand McNally's. Part of what makes this story of Agloe and its general store so appealing, and so endlessly re-tellable, is that there are no winners or losers. It's not really a cautionary tale because, somewhat ironically, no one made any mistakes. It's a classic farce in which each logical step leads to the logical next step with ultimately very silly consequences, and none of the parties involved seems to have done anything malicious at any time.

However . . .

'THERE WAS NEVER A GENERAL STORE'

There is a rival story, one that doesn't particularly clear up how real or not Agloe really was or isn't. Because, according to the family who owned the land the general store was supposedly on, it never actually existed.

In 2016 a woman called Darlene Beers, whose family owned the land that backed onto the site of so-called Agloe for generations,

---

[*] Normally used for writing very low 1950s prices that never go above $1.

told a journalist writing about the story for local paper the *Times Herald Record* that no one in her family remembered any 'Agloe General Store'. She said that in 1930 her grandfather Frank Nead had sold the land – which contained nothing but a small fishing lodge – for the measly price of $1 to an organisation going under the very curious name 'Agloe Associates' – a fact that's backed up by the Delaware County records. Shortly after the purchase, the lodge was renamed 'Agloe Lodge'.

That a business could be trading under this name *before* purchasing the land it's apparently named after raises no small amount of suspicion. The Nead family had long believed that the so-called 'Agloe Associates' was very likely a front for Rand McNally, who, realising they were about to be rumbled for copying Esso's map, preferred to hastily set up a phoney company and pay $1 for a fishing lodge rather than potentially millions of dollars in legal fees.

A bold claim indeed, but if true it would certainly explain why so many details about the innocent 'General Store', including its precise location, what it actually looked like, who owned it, when it opened, when it closed and why the building needed to be demolished without trace after closing remain so hazy. Could it be that the legendary story of the Agloe General Store, so faithfully retold again and again, is, in itself, a trap?

AGLOE'S LEGACY

Having burst fleetingly into bonafide existence, Agloe quickly returned once again to its original status – a paper town. But its presence on maps continued to have the mildest of consequences when in 1957 an article in the *New York Times* detailing 'scenic drives through the Catskills' recommended an 'unmarked country road that goes through Rockland and Agloe'.

In 1992, Otto's beloved General Drafting was absorbed by the American Map Company, which means the data for Agloe was passed down and down right into the digital age. Agloe was still there on the last American Map issue of *New York State* for Exxon in 1998, and it even showed up on Google Maps as recently as 2014. Google only deleted the mistake when an article about it appeared on a blog called *Strange Maps* written by Frank Jacobs, and within a week it finally disappeared from maps for the last time.

Today, the town of Agloe retains a cult following, and not just among map enthusiasts. It featured in John Green's 2008 teen novel *Paper Towns* and the 2015 teen film *Paper Towns*, generating two surges of awareness among teens, resulting in scores of poorly driven road trips as part of a pilgrimage to the remote spot. To make these trips a little less disappointing, somebody has nailed a tiny sign to a lamppost in more or less the right place that says, 'Welcome to Agloe – home of the Agloe General Store'. That sign is the only thing to be found in 'Agloe' today, and they've put it quite high up so you can't steal it (again), but it single-handedly ensures the continued existence of what may well be two false-hoods – the town of Agloe, and the story of its General Store.

There's a lesson here. Agloe became a real place just because people said it was one. It reminds us that maps are never fixed and that they have the power to make the world as much as represent it. It shows that shared belief alone, when it reaches a critical mass, really can conjure even the most fantastical concepts into what surely counts as existence, just like Santa Claus.*

As for copyright traps in maps, they're still very real indeed. There's no way to be completely sure where your nearest trap street is. But, as a fun exercise, why not get your comfy shoes on, grab your nearest map and go for a stomp around your local streets, avenues and alleyways, until you uncover one for yourself?

We'll tell you why not. Because it's a spectacular waste of your time. Tidy your room.

---

* If you're reading this with young children, go back and un-read that sentence.

8

THE MAP IN A BOX

Since first a map was ever made,
many years ago,
cartographers have drawn for us
the places that they know,
with so much detail on their maps –
however hard they try
some element or other
always seems to go awry.
From wonky borders, missing roads,
the absence of a key;
to surveyors in the field
getting high on mushroom tea;
the omission of a zero
can cause havoc with the scale,
though the good news is that none of these
will end you up in jail.

But don't be fooled, for map mistakes
aren't always so benign;
in certain countries certain maps
will see you doing time.
From publishing, distributing,
to hanging on your wall,
make sure your map is legal
(even if it's very small).
Cartographers are well aware
they must be diplomatic

'cos maps present a messy world
as simple and emphatic.
Their power can cause chaos
for an insecure dictator,
and certain maps can cause you
to be labelled as a traitor.
Though maps are flawed in many ways,
their strength can't be denied,
so mapmakers must make damn sure
their patron's satisfied.

For Chinese maps, be sure to check
the regions that they claim;
when choosing country colours,
make sure Taiwan's is the same.
Don't forget that in the sea
they'll want their nine-dash line,
and give 'Aksai Chin' to India
to land a hefty fine.
But then, if you're *in* India
they take a different view.
Aksai Chin belongs to them.
(The Depsang Plains do too.)
Which brings us to the conflict
with their smaller western neighbour.
If Kashmir's shown in Pakistan,
it's many years hard labour.*

In Argentina, maps of Argentina
have to show,
their claim to the Antarctic

---

* A caveat has been requested by our legal team. The punishment would not, of course, be nearly this extreme. Prison, yes, perhaps a fine, but nothing more (we checked). We may at times exaggerate for comedy's effect.

several miles down below.
Their constitution makes it clear
no map can be so mean as
to write the words 'The Falklands'
when they should be 'Las Malvinas'.

But what about the internet?
All good, you might assume.
Not so! Just 'cos your map's online
does not mean you're immune.

On Google Maps, for instance,
to avoid a legal sin,
their webpage varies greatly
based on where you're logging in.
If viewed from Turkey, Turkish Cyprus
(with which it's besotted)
is labelled with a solid line,
elsewhere it's shown as dotted.*

We look around the planet
at examples such as these,
and take comfort from the knowledge
we can publish as we please . . .
Or so we thought, but turns out
when it comes to these discussions
that even in the UK
there's a map with repercussions.

Let us take you on a journey
to some islands in the north,
a Scottish archipelago

---

* But if they happen to be using Surfshark VPN, they'll get to see the island with the dotted line again.

far from the Firth of Forth.
The islands known as 'Shetland'
can be found around halfway
between the Scottish mainland
and the east coast of Norway.
And though once ruled by Norsemen
they're now Scottish through and through,
(we'll avoid the hackneyed clichés
about deep-fried Irn-Bru).
We cannot overstate
that they're incredibly remote –
you can fly, but it's expensive
and it's fourteen hours by boat.

If you like big crowds of people
then a trip here is absurd,
though they're quite the party hot spot
for the migratory bird.
They're technically subarctic,
and they're windy, wet and cold,
so the kind of folk that live there
must be strong and tough and bold;
the kind of personality
to have a strong opinion
about the way that published maps
portray their wild dominion.
And as it just would happen
they had reason to complain
of the way that maps of Scotland
represented their terrain.
It wasn't misplaced towns or hills,
or poorly labelled lochs,
their issue was that Shetland
always turned up in a box.

That's right! Most maps of Britain
(for a reason we'll soon say)
placed Shetland in a quadrangle
just east of Sinclair's Bay.
'Twas mislocated, dragged down south,
a geogra-abberation,
to reduce the miles and miles of sea
on maps of this great nation.
On posters, textbooks, presentations,
Scottish ten pound notes,
Shetland always found itself
boxed in near John o' Groats.
(From time to time there came a map
that didn't use this square,
though usually 'cos poor old Shetland
wasn't even there.)

But Shetlanders were all agreed:
'The box should be forbade!
It's important our remoteness
be now properly displayed.'
They were sick of southern Scotland
getting all the map's attentions,
and demanded a more honest map
with Shetland-sized dimensions.

'Course, Shetland's not the only place
found in a polygon,
from Hawaii to Alaska;
San Marino to Hong Kong.
While the former hope that one day
they'll be boxless and set free,
at our current time of writing
there's no change in policy.
For the latter, though, their boxes

tend to biggen up their size,
so they like their special scale
and the status it implies.

In Shetland, meanwhile,
people still felt angry and demeaned,
until there came a chance for change
in May 2018.
An amendment to the Islands Bill
was coming into force,
MSP Tavish Scott (Lib Dem)
proposed a change of course.
'The islanders have had enough!'
he cried while red of face.
'It's time to unbox Shetland,
and to show it in its place.
If it's the last thing that I e'er get done,
before I reach retirement,
it's see this House put into law
my Shetland maps requirement.'*

But that was just the start of this
cartography kerfuffle,
as news of an unboxing then caused
something of a scuffle.
The Ordnance Survey waded in,
quite keen to have their say:
'Boxes may not be ideal,
but sometimes they should stay;
the isles are plus one hundred miles
from mainland's northern spot,
to include them there on every map

---

* Another note from legal here: we're not quoting directly, we're constrained
by rhyme and metre but the gist we state correctly.

can impact quite a lot.
For one it means a smaller scale
(imagine zooming out);
so you're forced to lose some detail,
which some maps are all about.
Between Shetland and the Highlands
is a massive ocean gap,
leaving Scotland's biggest cities
at the bottom of the map.
Plus, it's not like their enclosure
is a modern mapping fix,
that Shetland box is on a chart
from 1736!'

When this argument went public,
well, the press arrived in flocks,
with headlines reading:
'NOBODY PUTS SHETLAND IN A BOX!'
Of course, it's hardly life or death
but far as we're aware,
it's the biggest Shetland 'thing'
since that detective series there.
(A side note: something similar
from 13 years before,
when the BBC's new weather map
created a furore;
they put the country on a tilt,
to make it more 3D,
but England now looked bigger,
leaving Scotland somewhat wee.
And yes, the Scots were outraged,
and the storm just grew and grew,
until the Beeb said, 'Sorry'
and resumed their bird's-eye view.)

While boxes may look ugly,
one must also understand,
there's quite a lot to think about
when mapping lots of land.
What's important? Distance?
Detail? Relativity?
Who will even read this map?
and where should Shetland be?
Perhaps it's right that maps of Scotland
made by those in power
display the islands back up north
to stop things turning sour.
And yet on many types of map
the information's key,
and it's senseless to de-centre
Glasgow, Edinburgh, Dundee.

But in the end the change of law
officially was passed,
and Shetland would be shown
on its true latitude at last!
(Although, the rule does not apply
to likes of you and me,
its reach only extends to maps
by public agencies.)
Since Shetlanders complained
their frequent boxing-in was cruel,
the United Kingdom's won itself
a special mapping rule.
But, as far as we can see,
no one's yet been in the stocks
for doing the misdeed
of putting Shetland in a box.

9

THE PARANOID MAP

With hindsight, we see that the Soviet Union never had a chance
of world domination, but we didn't know that then.
Ken Follett, author

[Editor's note: Just to flag with you both – and it shouldn't be a problem – but we had an issue last year with one of our printing houses being hacked by a state-affiliated Russian nationalist group. We have tightened our security since then, so fingers crossed we won't have any issues.]

We will start with the good and move on to the bad.

THE GOOD

Russia is, and for a long time has been, very good at producing maps.

At various points in history, they have mastered:

- **Triangulation** – measuring distances and angles between landmarks to survey the land. ✓
- **Geodesy** – the science of accurately measuring the earth's shape and size, which is a) harder than it sounds because the

earth is not a perfect sphere, and therefore b) important if
you don't want your distances to be all wrong. ✓
- Aerial photography ✓
- Producing clear, accurate topographic maps – i.e. maps that
  show the main natural and human features of the landscape,
  including where it's more uppy or downy (topography), i.e.
  the one you take with you on a hike.* ✓
- **Doing all of this across the biggest country the world has
  ever seen.** ✓

Russia is a country with many frontiers, meaning, like Channel
5 in 2008, it has lots of neighbours. Some to the west, some to the
south, and, in the case of the Kaliningrad exclave, some to the east
and the north as well. It's a country of enormous regional power,
but also has lots of disparate populations within its territory to
look after. For all these reasons, it's always been in Russia's interests to master the art of mapmaking.

Peter the First was the first to make it a real priority, ensuring it
was a part of the army's core skillset and responsibility. Fascinating chap, Peter. Could be a chapter in his own right. But it turns
out we don't have time to cover everything that's ever happened in
the history of Russian cartography. Instead, we're going to fast-
forward from the 18th century to the 20th, to Lenin, because this
chapter is specifically interested in how Russia fared when it was
Soviet Russia, which was Lenin's 'thing'.

Initially, it fared very well. Under Lenin's New Economic Policy
(the most blatantly un-socialist programme of Lenin's tenure, and
awkwardly the most economically successful), mapmaking flour-
ished. The Bolsheviks (Lenin's posse) wanted lots of excellent maps
of the USSR in lots of different scales, and they set about producing
them in the 1920s. A new mapping agency was created to produce
civil maps, alongside the military ones. Things were looking up.

---

* Preferably laminated, although for the fashion conscious a waterproof map
case with a drawstring around the neck will suffice.

Then, in 1924, Lenin died. And no sooner had he been preened, pickled and presented to the public, than someone else with quite a different leadership style took over.

THE BAD

In 1924 Joseph Stalin took charge of the Soviet Union.

Little changed for the first few years, but from about 1930 Stalin started to become more, for want of a better word . . . Staliny, and everything began to get worse, including for maps.

LÏƏSI ГĤÏS ÏS ДМЄЯÏÇДЙ РЯФРДGДЙDД. ЯЦSSÏД ЦДS ЙƏVƏЯ GФЙЄ ЬДÇКШДЯDS, ФЙLY ГФЯШДЯDS!

Controlling the maps

Under Lenin, some maps had been produced commercially by the Geodetic Administration and were funded by various organisations who ordered them. But Stalin, famously preferring the centralised approach to government, issued a series of decrees between 1931 and 1933 that started to bring mapmaking under control of the state – a sign of the totalitarian times to come.

In 1935 centralisation went a step further. All national mapmaking functions were to come under the direct supervision of the Narodnyy Komissariat Vnutrennikh Del, aka the People's Commissariat for Internal Affairs, aka the NKVD, aka the Soviet Secret Police, which tends not to be a wholly positive sign for any traditionally non-secret civil agency.

Worse still for fans of map consistency, no one in charge of this organisation managed to stay in the job for long. The chief of the NKVD, G. G. Yagoda, was ousted in 1936 and executed as an 'enemy of the people'. His successor was Nikolai Yezhov, nicknamed the 'Bloody Dwarf'. He too was ousted, and shot on the orders of the head of the secret police three years later in 1940.

Such issues with staff turnover caused the Great Soviet Mapping Exercise started under Lenin to slow dramatically.

State censorship, meanwhile, was picking up the pace. The secret police were gleefully issuing strict instructions on the production of maps, which were not to be deviated from under threat of jail/Gulag.*

This became a huge problem for anyone whose job involved producing any kind of map for any kind of purpose, such as for research. For example, if you'd chosen the high life and were working as a soil scientist, desperately trying to improve and expand agricultural efficiency across a country where famine was beginning to become an issue, you were sadly not allowed to store your own soil maps, making said critical job substantially harder.

Even weather maps were affected. Under Stalin's strict censorship, certain types of weather were seemingly banned from maps. As one scientist commented in his diary in 1941:

> A curious feature of our times is the unexpected and incomprehensible organised ignorance. [. . .] Not only are the maps no longer printed, but even [the mentions] of cyclones and anticyclones have disappeared. ÏЙ ЯЦSSЇД, ТНЭ SЦЙ ЇS ДLШДЧS SHЇЙЇИG

This, it turned out, was just the start. The leash around map production and publication was about to get a whole lot tighter, and more twisted, due to a somewhat significant conflict. In keeping with the contradictory nature of the Soviet Union in general, it was the high quality of their maps that ultimately led to their own downfall.

Second World War

On the eve of the war, Soviet cartography may have been heavily centralised and frustratingly administered, but it was also:

---

* Stalin's brutal forced-labour camps.

- Producing good, detailed, topographic maps in a variety of scales of Soviet territory.
- Winning international awards for its excellence.

Then in June 1941, two years into the war, the largest ever invading force to invade invaded the USSR. Stalin hadn't thought this was about to happen, but it did happen, and when it did, there was no denying it had. Operation Barbarossa, as it was known, was a German military operation that saw 3,000 tanks, 7,000 artillery pieces and 2,500 aircraft support a land invasion of Soviet territory.

Initial progress was – from a Soviet perspective – terrifyingly swift, as German forces swept across the western plains and rapidly advanced on Moscow. Millions of Red Army soldiers were killed and wounded, with millions more civilians starved, as Hitler greedily sent as much food back home to Germany as he could dig up along the way.

On reaching the outskirts of Moscow, however, German progress stalled. Russia's greatest natural weapon, winter, eventually came to Stalin's aid, forcing the Germans to retreat until the following year (when they would try and fail once again, in much the same manner).

The German invasion had ended in victory for the USSR by the skin of its teeth. When the smoke had cleared and guns were back in their boxes, Stalin was left asking, 'How on earth did the Germans get to Moscow so quickly?'

Rather than blame his own refusal to countenance the possibility of the German invasion (which British intelligence had forewarned him about and he'd chosen not to believe), or the fact that he'd purged many of his most competent military leaders in the late 1930s, Stalin pointed the finger at what he thought was the obvious culprit:

'Our maps are too good!' he whined. 'We're too *capable*.'

He had a point. ⌊ТНДЙК ҰФЦ‼ ⌋ It was believed that in the early stages of the war, German intelligence had hoovered up the new Soviet topographic maps, giving them an enormous leg-up in reaching

Moscow as quickly and brutally as they had. Key transportation routes, industrial sites and agricultural centres had all been targeted and crippled by Hitler's forces during the campaign, presumably with the help of the impressively detailed and accurate Soviet maps.

It was clear to Stalin a change of approach was needed.

Maps become contraband

The repercussions of the invasion fiasco were severe for both mapmakers and for maps. Topographic maps had always been classified by the state, only to be used by the military or specific civil agencies who needed them, but the scale of the censorship was now to be stepped up a notch.

From 1942, large-scale maps of Moscow were banned from public circulation, removed from libraries and even burnt. The international award-winning Soviet *Atlas of the World* was withdrawn from circulation, thematic maps that showed useful information such as levels of industrial production were classified for the next fifty or so years, and ████████████████████ █████████████████, hilariously.

To be fair to Joe, this sort of strategic map censorship wasn't exclusive to the Soviet Union. In fact, preventing your enemy from getting its hands on maps has been a tactic used throughout history by anyone trying to maintain a strategic advantage over a rival. Napoleon, for instance, was obsessed with maps, commissioning them in great numbers and jealously guarding their secrets as he racked up his countless military victories with the aid of the exclusive strategic advantages they provided. The Portuguese, for also instance, had sailed the globe centuries earlier charting foreign lands and 'discoveries'. Such was their enormous currency to the state that anyone caught copying sea maps of the coast of Africa faced torture and death.

This practice of suppressing sensitive information on maps continues to this day. Following the 9/11 attacks on the World Trade

Center and the Pentagon in 2001, rooftop details on a number of public buildings were blanked out on publicly available online aerial photography, for fear of similar attacks being carried out thereafter.*

So, the Soviets' decision to restrict the availability of detailed maps while there was a massive war going on was par for the course and, frankly, completely sensible. But what made their map-censorship particularly comment-worthy was its absurdity and pointlessness during the next significant chapter in its history: the Cold War.

The Cold War

What began as a tussle between the two superpowers of the mid-20th century – the Soviet Union and the USA – spiralled into a uranium-fuelled ideological stand-off that both dragged in and gave nightmares to the entire rest of the world.

By the start of the Cold War, literally all topographic maps in the USSR were marked: 'Secret – for official use only'. If you wanted to go on a hike in the Urals and get home safely, you were well advised to take a ball of twine with you. The Soviets were still making detailed topographic maps, of course. Lots of them, in fact. But most people weren't allowed to look at them. Which is a shame, because their production came at great financial and human cost.

The surveying parties dispatched across the cold and extreme expanses of Soviet territory suffered a considerable death toll as a result of starvation, cold and natural predators. A surveyor turned academic named Alexey Postnikov, who himself survived this brutal mapmaking ordeal, recalled stumbling across the traces of

* You can see some censorship for yourself on Google Earth right now. It's standard practice for militarily sensitive areas to appear as a blur, although ironically – in a sort of state-level Streisand effect – this has the self-defeating effect of making people who didn't previously know there was anything sensitive there super-curious and far more likely to go a-snooping.

former surveying parties on a regular basis. On one occasion in 1963, he found a camp in the Uchur valley in the far east of the country that had been destroyed by bears. A note, dating from November 1948, was pinned on a tree trunk:

> All my reindeer had [sic] perished. The Yakut deer herd had died. There is no more ammunition. The food stores became bears' prey. I am left with a very sick junior surveyor on my hands. I have no transportation or means of subsistence. I shall try to force my way to the River Gynym.

Given he was 200 kilometres from the River Gynym, Alexey deemed the man's survival unlikely.

Just a handful of top brass were allowed to admire and utilise the fruits of the surveyors' sacrifices. Meanwhile, the only topographic maps available to the general public in the Soviet Union were at a scale of 1:2,500,000,* which were as handy as using a map of the entire UK from the front cover of a road atlas in the glovebox of the car to locate the Red Lion pub in Kidderminster.

Of course, other non-topographic maps were also still needed and circulated, such as tourist maps, road maps and maps for use in educational resources. These maps wouldn't typically display sensitive military or industrial information, but that didn't stop the paranoid authorities from closely monitoring and regulating them too.

Road maps, for example, failed to differentiate different types of road. All roads, whatever their surface, were simply labelled as a generic 'road'. Was it paved? Was it a dirt-track? The only way to find out was the hard way. Navigation was hard enough for people who lived in these places, but spare a thought for any visitors to the

---

* In 1965 even the 1:2,500,000 maps were withdrawn from circulation. This was particularly embarrassing, as the Soviet Union had been collaborating with its Warsaw Pact allies on a world map at this scale. According to its own rules, the USSR itself could now no longer feature on the map.

region, as tourist maps were just as unhelpful. Once again, only small scales were used (i.e. zoomed out), meaning they were broadly sketched and lacking in detail, often leaving only the most skeletal outline of an area to go by. We've reproduced two maps of the same area of Riga, Latvia (part of the USSR at the time) that show just how easy it would have been to get lost as a tourist here during the Cold War era. On the left is a c.1980 Soviet tourist map of the city; on the right, a part of the same city on a map published after 1991 following the country's independence. At first glance, it might look like the one on the left is simply at a different scale/level of detail compared with the one on the right, but look more carefully. It isn't just less detailed – it's plain wrong. Compare the land around the canal on each, and you'll see exactly how many streets and buildings the Soviet map feels happy to airbrush out of existence.

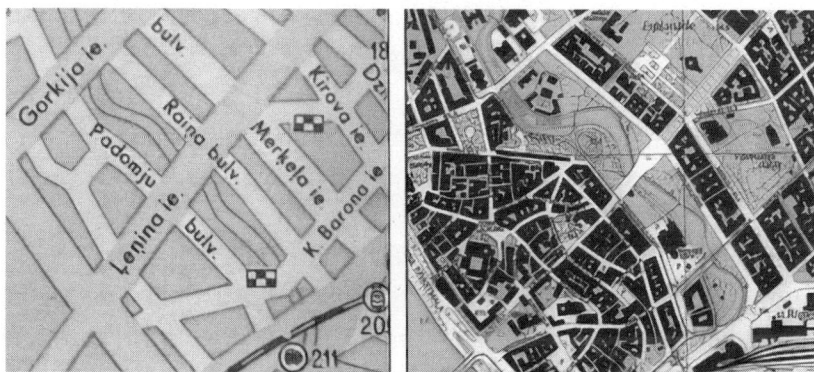

Courtesy Dr Alex Kent.

The irony here is the newer map on the right, while published after the collapse of the USSR, was nonetheless based partly on old USSR-era map data, kept secret from the general public. The information was always there, just not for people to actually see or use. But in one sense tourists to Riga could count themselves lucky. It's been estimated that due to cartographic secrecy and paranoia during the Cold War, around 30 entire cities in the Soviet Union simply vanished from maps altogether.

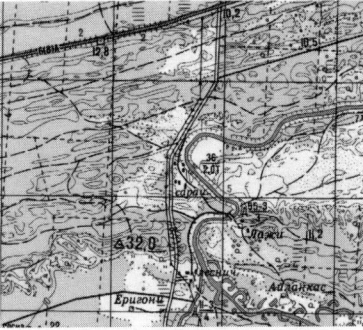

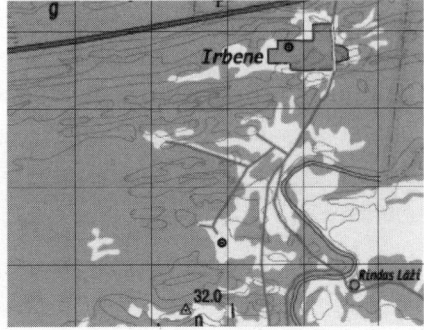

Left: Soviet 1:50,000 military map published in 1988 (no Irbene).
Right: Extract from 1:16,620 map of Latvia, published in
2010 (with Irbene).

One such disappearance can be found elsewhere in Latvia.* The town of Irbene on its Baltic coast was built in 1971, primarily to house a secret radio telescope. But it was also a fully functioning town with accommodation, a school and concert hall, all for the use of the military officers staffing the telescope. Quite how any visiting orchestras found it, however, is another matter, as it was never actually shown on any Soviet maps, the only clue to its existence being an unassuming white splodge in the middle of some green (in the original colour version).

Irbene was abandoned following the withdrawal of the Soviet army in 1993, but the derelict buildings still survive (as does the telescope). Despite it now *not* being a town, it now *does* feature on most maps of the region. Strange world.

* The reason for Latvia's recurrence in this chapter is because there's no complete or easily accessible archive of Soviet-era maps, meaning that the discovery of a huge trove of them in a Latvian map shop in the early 2000s by author and map-enthusiast John Davies was enormously useful. By happy coincidence, Dr Alex Kent discovered a separate haul of Soviet topographic maps in a map shop in Kazakhstan, thousands of kilometres away. John and Alex then found each other and joined forces to produce an excellent book about how the USSR mapped huge swathes of the world during the Cold War in incredible detail without anyone noticing called 'The Red Atlas'. Naturally, we recommend.

Another method of ensuring published maps could meet the necessary security guidelines was to only print an oddly limited section of the region in question, yet still try to make it fit for purpose. One example is route maps that take the form of a strip. These would only show the route you would take from, say, Moscow to Stalingrad, and none of the surrounding countryside or towns along the way, much like the strip maps for pilgrims used in Europe in the 1300s (more on these in Chapter 16).

At times, the censorship bordered on the absurd. One cartographer in communist East Germany recalled wanting to publish a map with a section of Finnish Lapland. This appeared uncontroversial; Finland was not a part of Soviet-controlled territory, so it shouldn't have been a problem. Said cartographer completed the map, lined up the print order, sought the necessary permission from the Berlin censorship office to press print . . . and was refused permission, for the baffling reason that:

> The map would include contours, and since it was not desired that Finland publish contour maps of the GDR, such a map of Finland could not be published here either.

Major security breach averted, clearly.

The most striking example of map censorship, however, is detailed in a trove of never-before-seen documents – sealed 48 years ago – to which we've obtained exclusive access and can now reveal in full.

In 1976 the ███████ published a series of ███████ of the city of █████. They all featured ████████, such as ██████ ██████ and ████████, but in a bid to ████████████ each ██████ had been replaced with a ████████████. But one citizen by the name of ████████ looked at ████████ and realised what had been happening. He ████████ on the corner of ██ ██████ and ████ with a ████████ and ████████████ ████████████ and ████████████████████. The ████ ██████ concerned that his ████████████████████ immediately

██

████████████████.

Now, ██

██

██

██

██

████████ the ████████████████████████

████████ if ████████████████████████████

████████████████████████████████ He ████████

██

████████████████ on ████.

██

██

██

██

██

██

████████████████████████████ and ████████████

██

██

██

██

████████ oink! ████████████████████████████

████████████████████████████.

Then, ████████████████████████████████████

██, asking,

'██

████████████████████████████, which you can still see today if you look at any map of Rutland.

Second, ██████████████████████████████████

██

██

██

than

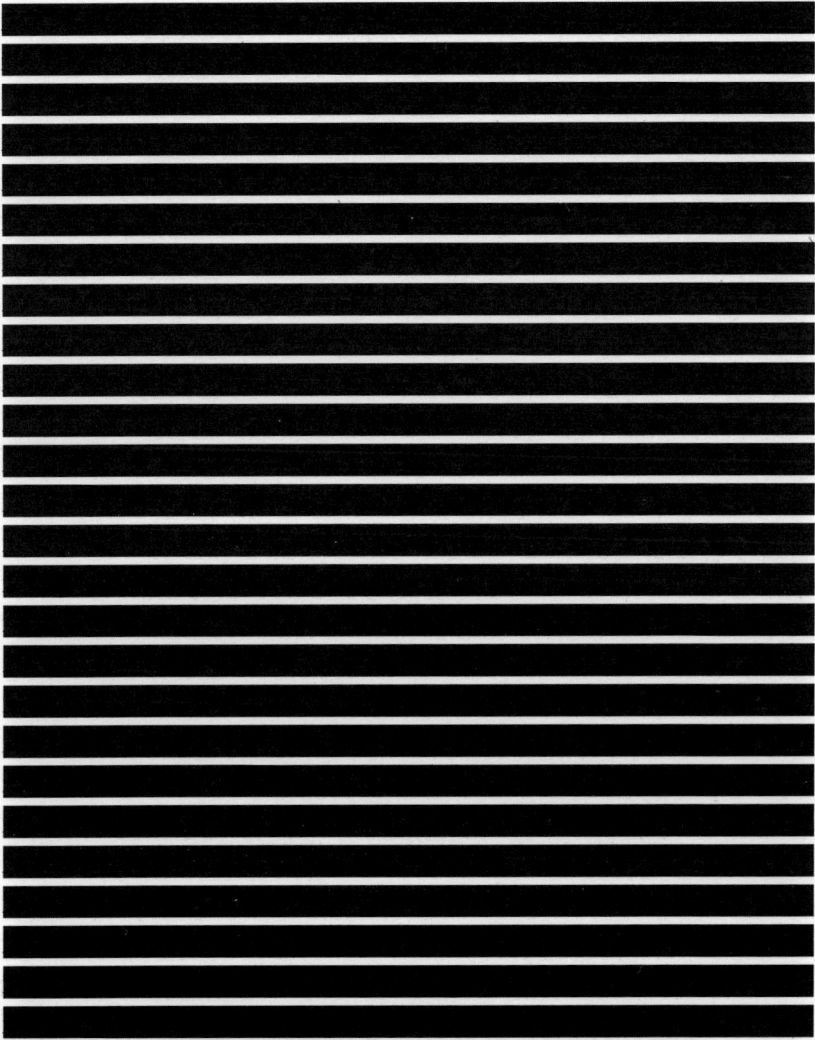

Blitzen the black-nosed reindeer, ♫ had a perfectly ordinary nose, ♫ and if you ever saw it, ♫ you wouldn't say anything at all, ♫ all of the other reindeer, ♫ treated him like everybody else

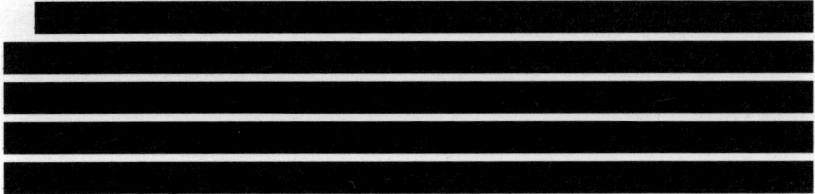

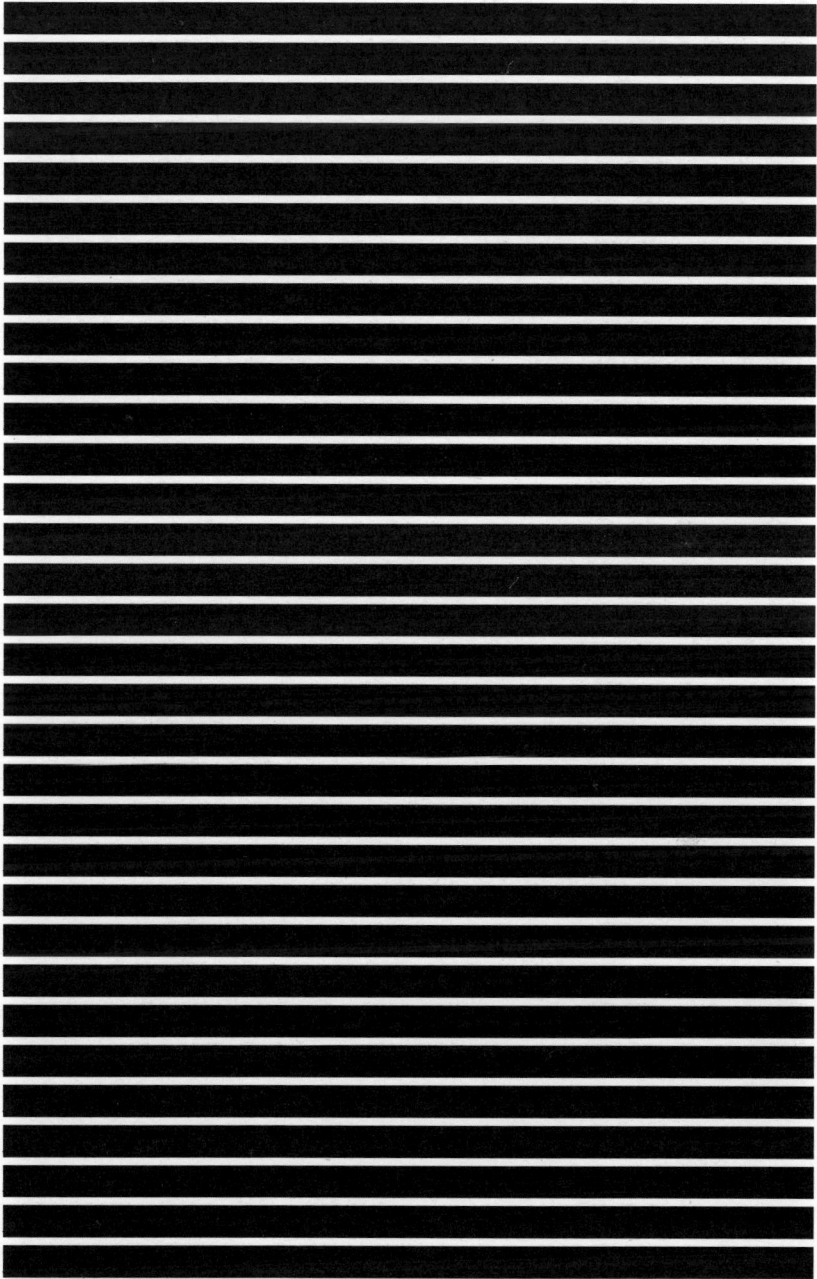

thus ending all life on earth on Wednesday 24 January 2029.

*

Censoring maps and restricting their distribution undoubtedly made the geographical dimensions of Soviet people's lives harder during the Cold War, but it was about to get a whole lot worse . . . (again). As the stand-off intensified, as a hot war began to feel ever closer and as the paranoia dial swung into the red, the authorities decided to turn all maps used by the public into a confounding hall of mirrors.

Intentional wrongness

Theodore Shabad was a map-enthusiast with a keen eye for detail. A reporter and editor at the *New York Times*, he also edited the journal *Soviet Geography* on the side, and he loved nothing more than to while away the hours studying maps.

One possibly sunny, possibly Sunday, possibly afternoon in nineteen sixty-something, he found himself comparing new Soviet atlases with older maps of the same territory when he began to notice something odd. Entire cities, roads, rivers and landmarks – whose locations had long been known and recorded – appeared to have moved on the new atlases, sometimes by small amounts, sometimes by as much as 200 miles. The more Shabad checked, the more discrepancies he found, including between different editions of the newer atlases. Soviet-built railways and frontier settlements were often not marked at all, despite their construction having been lauded upon completion. Towns would disappear then reappear in slightly different places across the different editions.

Theodore had inadvertently stumbled across a fact known to many Soviet citizens who'd consulted maps of their locality in recent years: that the USSR was no longer satisfied with a policy of map censorship. Now, mapmakers were intentionally misplacing a number of features on the assumption that foreigners like Theodore would get their hands on maps and end up being confused, which Theodore was. In truth, they were probably less concerned with Theodore and more concerned with any military top brass considering aiming ballistic weapons at the Motherland.

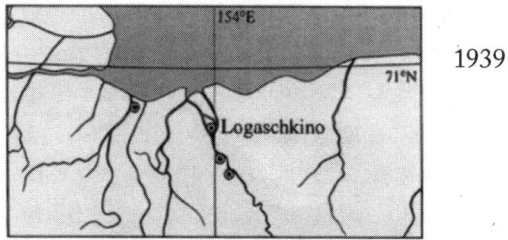

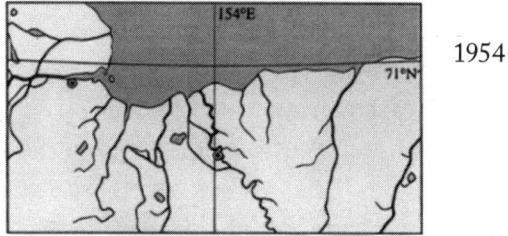

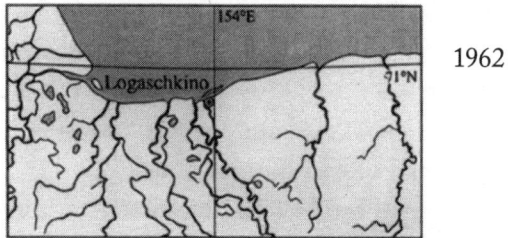

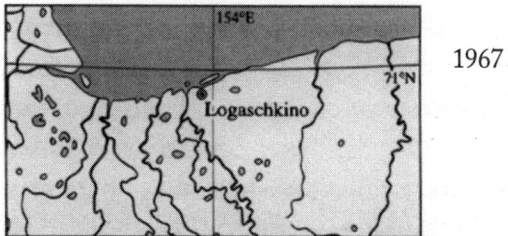

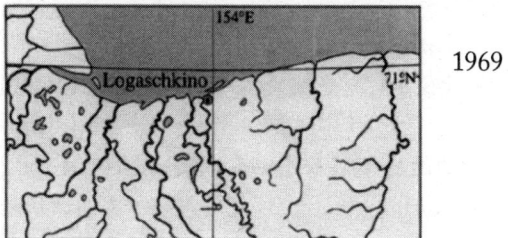

(Extracts from the AV map sheet Olbernhau (second edition, 1978, current to 1979) in the south-east of the GDR.)

Around 1970, US army officers began compiling a list of the 'puzzling peregrinations'* of towns on Soviet maps. One such town was the now abandoned town of Logashkino in the far north-east of the country. A small trading post in the Arctic tundra, surrounded by swamps and lakes, it was nonetheless deemed of enough significance to have its location altered (or disappeared) with each new edition of the great Soviet *World Atlas*.

The town appears, disappears, reappears on the coast at the end of a river, then the river disappears and the town moves further west of the 154° east meridian, before finally cosying up to that meridian once again in 1969. Some of these movements may appear so minor as to be of little consequence, but they become more significant when you consider the United States was developing a global geodetic coordinate system, a technology that could in theory allow electronically guided missiles to deliver nuclear warheads with unprecedented precision. Faced with this threat, Stalin and his successors decided it was 1) safer and 2) far more entertaining to hide potential military targets – just like a game of *Battleships* – and obscure their location. This meant that if the US *did* select a coordinate to fire at, the Soviets could then hop on the famous Red Telephone and tell them whether it was a 'hit' or a 'miss'.

In many cases, while Soviet authorities appeared to be making maps for the benefit of civil agencies, they were in fact producing them for the specific dis-benefit of foreign spies and military staff. And nowhere was this cartographic bluffing more pronounced than in the German Democratic Republic.

EAST GERMANY SEES DOUBLE

Following the end of the Second World War, East Germany – one of the USSR's satellite states – found itself at the business end of the Soviet Union's much prized 'buffer zone'. This was a huge swathe of

---

* Their words. We approve.

Eastern Europe, which, under Soviet control, formed its western frontier with the West and could help protect heartland cities like Moscow and St Petersburg from the invading armies of the future. If there was ever going to be a clash of land-based forces in the Cold War, the GDR was bound to play host to at least part of the conflict. As a result, protecting its territorial secrets by telling outrageous porkies on its maps became a key tool in its defence.

In the years after the war, East Germany was in desperate need of topographic maps to assist with the rebuilding of the country. Happily, these were readily available – until, that is, the entirely arbitrary year of 1965, when the ever-paranoid Soviet authorities decided that the time had suddenly come to make them all completely secret. Balancing the contradictory needs of letting the rebuilding continue and keeping their territory safe, they came up with A Very Soviet Solution.

The plan was to produce a second series of topographic maps of the entire territory, similar but noticeably different to the detailed and officially held restricted ones. This second set of maps would be made available to people outside the military – not just any old civilian punter, mind, only organisations and government agencies who needed them to do their job.

Even though recipients of these non-military maps would still require a clearance level to view them, the idea that there could be any sort of wider access to maps had the paranoia dial twitching once again in Moscow. This second set of maps – known as the 'Edition for the National Economy' – was therefore not allowed to be anywhere near as detailed or useful as the highly confidential official military ones. What's more, working on the assumption that these maps would undoubtedly end up in the hands of Western enemies, they needed to be purposefully wrong.

This wrongness took a number of forms, the most foundational being the concealment of the entire underlying grid system. In short, to convert the irregularly shaped, almost spherical earth onto a flat map, you need to do some maths. Not just deciding which projection you're using, but which geodetic datum you're starting from (the model that describes the shape and size of the earth's surface and provides a reference for your coordinates). Because nobody knew the underlying model used on the new GDR maps, it was impossible to deduce the actual co-ordinates of any location, a considerable problem if you needed to use a map for purposes that required accuracy – such as mining, laying rail tracks or building a thing in the actual right place.

Furthermore, where most topographic maps reference lines of latitude or longitude somewhere in the margins, here, none were marked. This meant the grid lines on the map referred to nothing other than each other. But this concealment of co-ordinates wasn't the sole extent of the confusion, not by a long chalk.

The positions of prominent landmarks such as churches, masts and buildings were frequently altered, and city centres were also displaced by up to three kilometres. Map symbols were recoded to deter potential armies from certain areas that might help them advance – newly planted forests, for instance, were indicated as being mature forests, major roads were regularly reassigned to the status of mere farm or forest lanes and larger building complexes were labelled as detached houses.

It's hard to understand why the authorities bothered putting the effort into producing these maps at all, especially when you consider that making this very similar but also frustratingly different set of maps across the entire GDR was cripplingly expensive to carry out, in a country that was in desperate need of cash.

And it wasn't just foreigners that Soviet authorities were concealing intelligence from on these secondary maps. Here in the GDR it was also the citizens themselves; folk who couldn't be trusted not to dream about the allure of West Germany and escaping across the border. WHY WOULD ANYONE 'ESCAPE' THE BEST QUALITY OF LIFE THE WORLD HAS EVER SEEN?

The below map, from the 'National Economy' edition of the Soviet GDR maps, shows a north-western border with the Federal Republic of Germany (the left bit is shown here in white because the West was not a region for people to be imagining in anything other than featureless solid monochrome).

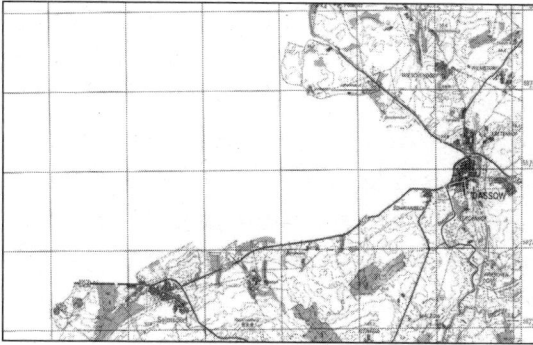

But this wasn't the actual location of the border with West Germany. The true border was up to three and a half kilometres away, as indicated on the below topographic Soviet map produced for the military.

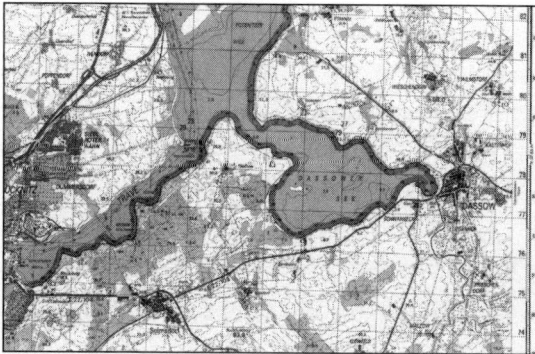

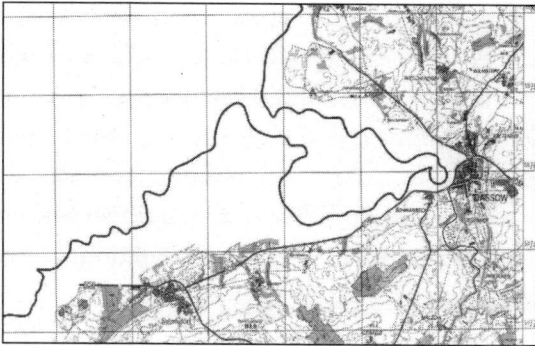

(Extracts from the AV map sheet Schönberg (the second edition, 1978, current to 1979) in the far north-west of the GDR.)

Transposing that border onto the first map, we can see the zone of GDR territory that has *also* been obscured by whiteness before the actual border is reached. But why?

It was assumed by the paranoid authorities that even employees in GDR businesses who were granted access to the 'National Economy' edition maps, and had therefore been security checked, might nonetheless still want to run away from the GDR. The best way of ensuring they didn't was to allow them to get lost in a big white nothing if they so much as tried.

The achievement of the Soviets' parallel mapping exercise in the GDR was accompanied by:

- an enormous cost of production
- misguided planning and project management for those using the maps
- stifling of infrastructure development
- a huge administrative effort devising the falsified data
- the anger and frustration of anyone who tried to use the maps, with a subsequent loss of Soviet political capital
- lost and confused citizens

Happily, however, for any behind-schedule Berlin builders, delayed Dresden diggers and late-running Leipzig labourers, a technological development on the other side of the planet would soon render this double-drawing exercise completely pointless.

Throughout the 1950s and 60s, the US military had been developing and mastering the art – and science – of space photography, enabling them to build up a clear picture of what was where anywhere in the world by zooming in from space. With increasingly detailed images from across Planet Earth, including every inch of Soviet territory, the US had all the information it needed to pinpoint what was really where on a flat map.*

---

* Incidentally, the Soviets had some pretty good satellite photography of their own, so they of all people should have known what the Americans had at their disposal.

But, absolutely bafflingly, despite knowing the Americans were armed with such comprehensive intelligence, the Soviet Union doggedly continued, at tremendous expense, to intentionally produce misleading maps for at least 15 more years, for the pointless bamboozlement of no one but their own citizens.

The absurdity could not last much longer.

THE TRUTH FINALLY COMES OUT

In 1988, as the Cold War was starting to draw to a close, the long-time propaganda mouthpiece of the Soviet state – *Izvestia* newspaper – published an interview with the chief of the USSR's principal mapping agency, Viktor Yashchenko, in which he spilled the cartographical beans, confessing everything.

Viktor admitted that the Soviet Union had deliberately falsified virtually all public maps of the country for 50 years – misplacing rivers and streets, distorting boundaries and omitting geographical features, on orders of the secret police.

Roads and rivers were moved . . . city districts were tilted. Streets and houses were incorrectly indicated. For example, on the tourist map of Moscow, only the contours of the capital are accurate . . . Tourists tried in vain to orient themselves.

He even conceded there were people clever enough to notice.

We received numerous complaints . . . People did not recognize their motherland on maps.

Perhaps the most revealing thing about life in the USSR was that when this information finally went public, no one in the country seemed surprised or bothered. After decades of coping with food shortages, strict censorship and propaganda, widespread surveillance and repression, limited personal freedoms and travel bans, miserable mapping was clearly the sort of thing they'd just come to expect from their government.

Viktor ended his interview with *Izvestia* with a radical promise:

From now on, all places, points and other information will be shown where they really are.

While that's a bit like a bus driver pledging that he's going to start driving a bus from now on, for the USSR, a mapmaker promising accurate maps was a pretty big step.

This interview marked a turning point in the cartographic culture of the USSR. Up until then, wrongness had been not only expected, but actively celebrated – one academic who devised a special cartographic projection that led to random distortions in coordinates was awarded a state prize. But this surprisingly candid interview was an early sign that things were about to change for the better for Soviet citizens. The USSR would only survive for another three years before ███████████████████████████ ███ ███████████ which is why the biggest part of the country is now Russia again.

It's become a cliché to describe the Soviet Union using words such as 'paradox', 'contradiction' and 'irony'. But the story of these falsified, distorted and mad maps shows the Soviet Union was an ironic contradictory paradox.

The mapmakers of the Soviet Union in the second half of the 20th century produced some of the best, most accurate, and, to boot, most stylish topographic maps the world had ever seen.

It's just a pity hardly anyone saw them.

10

THE DEADLY SHORTCUT

LEFT, FOR DEAD

Podcast transcript
(w/full audio description)
2 × 30–minute episodes
Presented by Caleb Carter

Episode 1

MUSIC: a slow and melancholy melody played on a lone cello is joined by some light, sparse, high-pitched piano notes, adding mystery and intrigue to the bass note of gravity.

Slight music fade as the voice of Caleb Carter pushes through the enticing theme. Caleb has a strong New York accent and speaks with a confident self-assurance, as if he's really done his research. His voice conveys a studied honesty; you know he's not exaggerating the facts because he practically throws them away with an insouciant if not highly derivative vocal fry.

The music underscores the following introduction.

CALEB: What would it take for you to eat another human being?

SFX: frontier marching drums.

Extreme hunger? A vendetta? Curiosity?

Perhaps you think you would never, under any circumstances, eat someone. But then, can you really know that about yourself?

I'm not sure *I* can.

This is a story which, if you're from America, you've likely heard before. I'll never forget the spring day in fifth grade when Mrs Hoffman first scratched out the words 'The Donner Party' on the whiteboard, then underlined it with a flourish as she turned to throw the class a menacing look, challenging us not to be horrified by what we were about to learn.

Come to think of it . . . I don't think any of us was able to meet that challenge.

If you're *not* from America, chances are you've never heard this story. Perhaps the words 'Donner Party' sound like little more than a late-night drunken indulgence, where a group of friends stagger into a fast-food joint to gorge on thinly sliced strips of barely edible meat.

Funny thing is, if that *is* what you imagined, you'd have got it half right.

*   *   *   *

CALEB VOICEOVER: The story takes place over a hundred and fifty years ago, in 1846...

EXPERT (unknown female voice): These were some of the first Americans to emigrate to California, setting out with their young families across some of the country's most hostile environments in search of a better life.

CALEB VOICEOVER: . . . a time not just before road signs, but, in much of America, before roads themselves, when these early pioneer families were forced to haul wagons across rivers and through mountain ranges on a treacherous 2,000-mile journey. And the best route to take? Well, it wasn't always obvious.

CALEB INTERVIEWING: What *was* the Hastings Cutoff?

EXPERT: It was supposed to be a new shortcut to California – but for those who risked it, it proved to be a disaster . . .

CALEB VOICEOVER: . . . and where a party of 81 people became stranded in the mountains over winter, resorting to desperate measures to survive. All because of one, deadly decision.
They turned left.

MUSIC: theme surges.

It's hard to forget the gruesome details of what came next, as we're reminded via semi-frequent pop-culture references. Heck, it even featured in *The Simpsons*.

But for my part, I realised I still had questions about what actually happened; ones that could only be satisfied by presenting a long-form podcast, rather than, say, simply reading about it in my own time. Questions like, how can nature be so beautiful, and yet also so deadly? How could a simple wrong turn lead to such unthinkable tragedy? And how could I avoid sensationalising the cannibalism at the heart of this story in a way that would help the podcast win a prestigious Ambie award, while also ensuring it featured heavily enough in the marketing to do well in the charts.

When I started my journey, none of these questions had answers. Then one day, the phone rang.

SFX: phone rings.

FEMALE CALLER: Hello, Mr Carter? Congratulations, you've just won $15,000 in our prize draw, would you like to claim your . . .?

SFX: dial tone.

CALEB VOICEOVER: Right after that call, I began my research. This is Left, for Dead, Episode 1, The Shortcut

MUSIC: the theme rises to a stirring musical climax, then abruptly cuts.

*　　*　　*　　*

SFX: a howling wind.

CALEB: 2 a.m. on September 4, 1846. A terrified family of six huddles, alone and unsheltered, in the middle of the freezing, lifeless salt flats of what it is today Utah. James Reed and his wife, Margret, sit upwind of their four young children – Virginia, aged 13, Patty, aged eight, James, aged five, and Thomas, just three years old – taking the full force of a howling gale to protect their children from the cold, praying that their desperately thirsty family makes it through the perilous night and that help is soon on its way.

The Reeds are just one of the families in a larger group of emigrants known as the Donner Party – named after their elected leader, George Donner – who are now four months into their long journey from Missouri to California. For the last three days they have been struggling across the waterless desert. Some have become so thirsty they've begun to hallucinate, the livestock are falling . . .

SFX: a single loud wail of an ox.

. . . and the party, usually travelling as one, has splintered in a race to reach water on the other side, with the Reeds at the back.

For thirteen-year-old Virginia, writing years later, 'It seemed as though the hand of death had been laid upon the country.'

The Reeds' animals had fared worse than others, so James had instructed his teamsters to unhitch the ailing livestock from the wagons and drive them on to water. Once replenished, they'd then return to collect him, his wife and their four weakened children as soon as they could. That was more than a day ago. Now in their second night alone on the flats, they wondered if help would ever reach them.

Facing both the bitter wind and his own mortality, James Reed might well have pondered his decision to personally persuade the party to gamble on the shortcut that had led them onto the salt flats, following an untested route that had been advertised by a man called Lansford W. Hastings.

As part of Hastings's sales pitch, he'd been led to believe the challenging salt desert was a 40-mile march. The reality was double that, pitching tired and vulnerable young families against the most unforgiving of landscapes.

If James regretted his gamble, he never admitted as much in his diaries. Reed was a sort of blustering anti-hero with a giant ego. I guess you could say he had 'main character energy', meaning he often rubbed people up the wrong way. He'd made his money in mining and was the kind of guy who couldn't help but boast about it. A mixture of vanity and charisma had made him de facto leader of the party, even if the quieter, more likable George Donner was officially in charge.

But for all his faults, James Reed was brave, defiant, and fiercely protective of his family. He and Margret clung on to their children for dear life, willing every long minute of the freezing night to pass.

When morning broke, the children all miraculously stirred as the rising sun's warmth thawed their bones. Eager to keep moving, James urged his exhausted family to walk onwards as he scanned the horizon for help. They soon reached the Donner family, who were sitting with their caravans, also waiting for help. Here, James decided to leave his family with the Donners, and march ahead to the spring at the base of Pilot Peak, where he personally rallied a group of men and livestock to find and save his family. (His own teamsters, it turned out, had been unable to help, as their livestock had bolted the moment they were unhitched from the carts.)

At nightfall on September 4, a full five and a half days since the Reed family had last encountered fresh water, James returned to escort the last members of the Donner Party safely across the salt flats.

Reed would have been forgiven for believing that the worst of the journey was now behind him, but he could never have imagined the nightmare that was still to come, nor the far more perilous rescue he would have to brave to save his family six months later.

Because the shortcut Reed had been so determined to take would cost them so much *lost* time, they were soon forced to navigate winter in the deadliest of environments.

*   *   *   *

SFX: typing and mouse clicking.

CALEB IN OFFICE: What's this? New email from my producer, Ava. Okay . . .

CALEB VOICEOVER (recorded in studio): Go take a look at your nearest map of America. Perhaps it's on your phone or framed on a wall nearby. Chances are there's a strong outline showing its edges and borders. And inside? Maybe the state lines are marked, or some highways linking up the major cities . . .

CALEB IN OFFICE: Open attachment . . .

CALEB VOICEOVER: And it's a funny thing, but when I look at a map like that, it seems to give these features a kind of permanence. An undeniability about their existence. As if Wyoming were just the most natural thing in the world; a perfect square predestined to delineate state law.

Now, all of a sudden, I was looking at a very different type of map.

CALEB IN OFFICE: 'John C. Frémont – map of exploring expedition . . . Oregon and North California 1843 . . .'.

SFX: clicking. Slight pause.

I can't . . . I mean what is this? Is this a virus? I'm calling Ava.

CALEB VOICEOVER: Because in 1846, when the Donner Party set off for California, most of the man-made features we're familiar with simply didn't exist. Most of America was completely *wild*. We're talking rivers, grassland, mountains, desert – and enormous unmapped areas.

An America I neither recognised nor understood.

CALEB IN OFFICE: Hi, Ava – what the hell is this attachment? It makes no sense.

CALEB VOICEOVER: So you know what I'm talking about, I've put a copy of the map Ava sent me on my website – www.calebcarterpresents.com. Check it out.

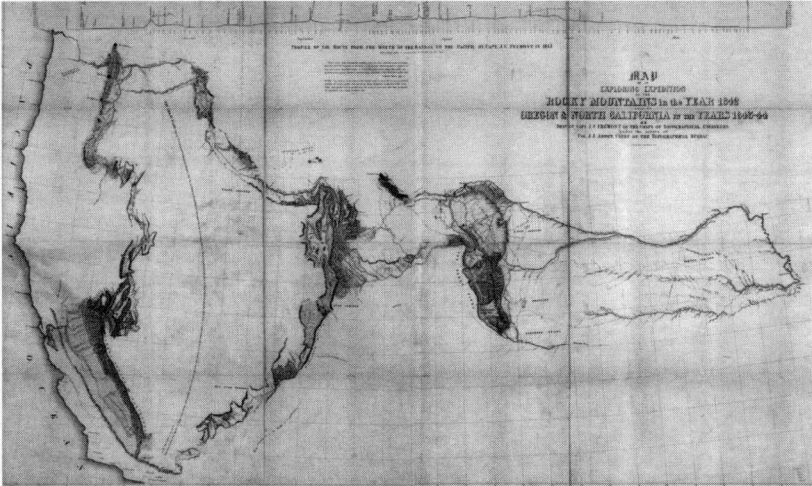

(*Courtesy calebcarterpresents.com.*)

AVA (down phone): Oh, OK. So, super-exciting – it's a map I found that was published the year before the Donner Party set off. It shows the route emigrants took from Missouri to Oregon and California!

CALEB: Even though she was young and naïve, I'd hired Ava because she seemed like a hard worker. Now, however, I was worried she might be completely mad. The twisted image in front of me looked more like a monkey swinging on a branch than a map of America, let alone one you could use to navigate 2,000 miles.

But before completely dismissing it, I decided I should speak to an expert in case I was missing something about this simian artefact, someone who would realistically agree to give up their time for a podcast that didn't yet have a broadcast deal and wasn't really paying their staff, let alone their contributors.

Luckily, I knew just the person.

*   *   *   *

CALEB INTERVIEWING: I gotta say, Mrs Hoffman, the last time I saw you I think you were storming out of our eighth-grade class because we were 'unteachable'? Was that it?

MRS HOFFMAN (the 'expert' voice from the introduction): *You* were unteachable, Carter. Singular. And let me be clear, I'm not doing this for you. I'm doing this because I have a PhD in early American history and have been consigned to dumbing it down for delinquents like you my whole life. Cut the chit-chat, and I'll tell you what I know.

CALEB VOICEOVER: This was classic Mrs Hoffman. She might now have looked decades older, but she still had her trademark wit.

CALEB INTERVIEWING: Got you! So, to start, I wondered if you could take a look at this map?

MRS HOFFMAN: Ah, yes. So this is a map of . . .

CALEB VOICEOVER: Mrs Hoffman had a lot to say, but I'm going to paraphrase most of it because her delivery . . . well, it was a tad flat. We'll come back to her for occasional quotes to remind you the interview really happened, but we'll keep them short.

She told me my producer, Ava, was right. This *was* a map of America – not a monkey at all – and it showed just how little of America had been mapped at the time the Donner Party set out. She said it was possible that some members of the Donner Party would have seen this map, or one like it, and would have used it to help guide their route to California.

CALEB INTERVIEWING: But how on earth could you get from Missouri to California with just *this*?

MRS HOFFMAN: It was no mean feat, and everyone in the Donner Party would have known they were taking a huge risk.

CALEB VOICEOVER: She said that besides maps like this, and descriptions of the route printed in pamphlets, emigrants relied on following the course of rivers, and the tracks of other parties who were ahead of them in the emigrant train.

But many felt the long journey would be worth it to reach California, which people back then imagined as a kind of paradise, even though Malibu hadn't even been built yet. Some were tempted by the climate or just a fresh start, others by the adventure of reaching the Pacific coast. And in doing so? . . . Fulfilling America's 'Manifest Destiny'.

By the way, I should mention that I know that at that time Native Americans already lived in what we now call 'America'. I just want to make sure you know I know that. They were good people, and we shouldn't erase their history. It's super-important.

Anyway, I still couldn't actually read this map, and knew I needed to understand the route itself. Where did it go, and what did it look like all those years ago?

Helpfully, Mrs Hoffman was able to show me another map from a textbook she brought from home. Again, you can check it out on my website.

She explained that the Donner and Reed families were both from Springfield in Illinois, but the journey really began when they got to Independence, Missouri. Apparently, back then, Independence was the frontier town that marked the edge of the abyss for civilisation. To the west, there was nothing but *nature*. Not even a whatever the equivalent of a 7-Eleven was back then, just the occasional 'fort' – usually just a small hut used by fur traders.

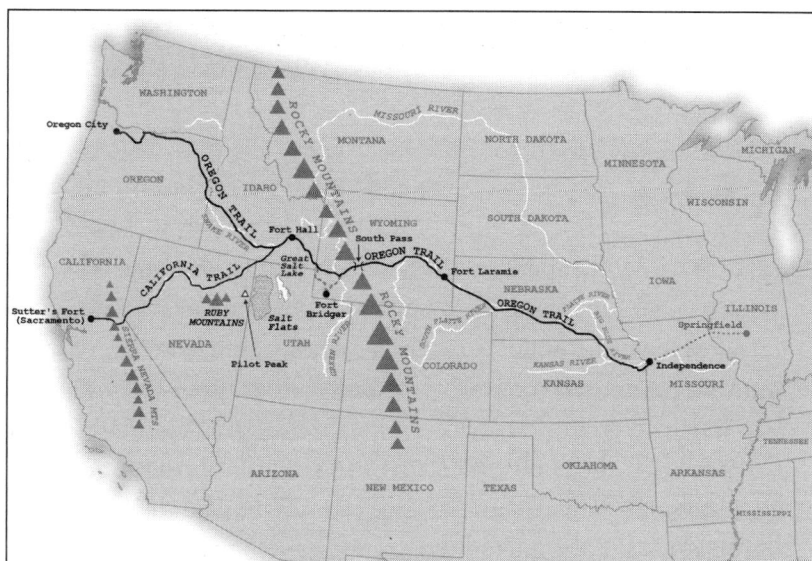

(*Courtesy calebcarterpresents.com.*)

MRS HOFFMAN: From Independence, most emigrants took the
Oregon Trail . . .

CALEB VOICEOVER: She said that from Independence, most
emigrants took the Oregon Trail, which followed a series of rivers
north-west to a low-lying crossing through the Rocky Mountains
called 'South Pass'. A little way after this came a fork in the road
where some would continue north-west to Oregon, and others
would head south-west to California.

The quickest you could hope to do the journey from
Independence to the west coast was four months if you were lucky,
partly because oxen can only cover 20 miles a day, at *best*. The trick
was to leave Missouri late enough in the year that there would be
plenty of fresh grass for the animals, but early enough to have cleared
the Sierra Nevada mountains on the eastern edge of California before
winter. That meant the best time to leave Missouri was mid-April.

But, for some reason, the Donners and Reeds only *got* to Missouri
in early May and set off from Independence on the 10th. At that time
nobody seemed too concerned. But according to Mrs Hoffman?

MRS HOFFMAN: They would have known they needed to hurry. And talk on the trail of a new shortcut would definitely have pricked their ears.

*   *   *   *

CALEB: It's funny, I've spent so much time getting to know the different characters of the Donner Party recently – learning the intimate details of their lives and personalities from the excessive summary notes my producer emails me – that they feel strangely like my friends.

There were a host of families that had banded together, not just the Reeds and Donners, but the large clans of the Murphys, Eddys and Breens – all with young children, and all with their own unique parts to play in the story. But it's the Reeds I've always felt particularly drawn to. My grandmother's maiden name was Reed, and I always wondered if I was related to that same family who set out in 1846 in search of a better life.

And sure, I know there's plenty of Reeds out there, but I've just got this feeling about it; like I've always been haunted by the story, knowing that there was more to discover and that I was the one to do it through my very own award-winning podcast.

Of course, I had to know for sure. So a few weeks back I sent off some saliva to ancestry.com. I haven't heard back yet – I'm actually worried I sent way too much saliva.

Anyway, back to the story.

*   *   *   *

SFX: a rushing river torrent.

CALEB VOICEOVER: A couple of weeks after leaving Independence, the Donners arrived at a known fording point across the Big Blue River in northern Kansas, but in front of them was an uncrossable raging torrent. Heavy rains had only just made the ford impassable –

they could see the party ahead pushing through the countryside having recently crossed. A storm broke ...

SFX: thunder.

... and they knew they'd have to wait days for the waters to subside.

That's right. They had to stop the journey, because of a *river*. I'd never really thought about rivers as an inconvenience before, so hearing this part of the story kind of blew my mind.

To me, bridges have always felt like one of those things that are completely natural. Like you see them in pictures of rivers all the time, and they just look like part of the landscape. And I know it seems obvious now I think about, but it never occurred to me that someone had built these bridges, and that there must have been a time when they simply weren't there. But it's true. They weren't.

In 1846, they *had* bridges, but not always exactly where you needed them.

CALEB INTERVIEWING: So how bad was this hold-up?

MRS HOFFMAN: On this journey, every day was crucial. Not just because they had to reach California before winter, but for supplies, energy levels and morale. Here beside the Big Blue River, morale was about to take a nosedive, because it was here the first member of the party passed away.

Caleb gasps.

CALEB VOICEOVER: Even though I already knew this, I was shocked.

Sarah Keyes was the elderly mother of Margret Reed (James's wife). To most, it had been clear from the start that she was too weak to make the journey. They think she died of consumption, which is kind of gross. Like, especially when food needed to be rationed and shared on such a long journey.

But I can't imagine how hard that must have been when Sarah died, because the men had to literally cut down a tree and build her a coffin. That would take me a year at least.

It was five long days until the party managed to cross the river on makeshift rafts, ready to continue their journey once more. They were now way at the back of the year's emigrant train, and in desperate need of a boost to help them make up the time. That desperation was to prove deadly.

*   *   *   *

CALEB VOICEOVER: We've all felt desperate before, for all kinds of reasons. But in my case, never more so than when I've had head lice. So that's why I'm grateful to our sponsor, Hank's Handy Head Louse Cream.

Hank's Handy Head Louse Cream uses the same active ingredient as an aggressive agricultural pesticide that's banned in most of Europe. So I mean it when I say the effects... are devastating.

Simply pour a blob onto the palm of your hand, rub it vigorously through your hair on the hour every hour, and in just a few days your hair will be totally head-louse-free.

And it's not just curative, it's *preventative* too. With my new shiny scalp I never have to worry about head lice ever again.

Simply sign up with the code OWWWWWWW for 20 per cent off. Hank's Handy Head Louse Cream. Wow! That's better. . .?

*   *   *   *

SFX: *muted indoor sounds, soft footsteps.*

CALEB (speaking in a hushed whisper): OK just through here? Great, thanks.

SFX: *door opening.*

CALEB: OK, wow! So, I'm in a medium-sized room in the archives of the Library of Congress in Washington, DC, and I'm here to view a very special document. It's called the *Emigrants' Guide to Oregon and California*, and it was written by Lansford W. Hastings, a man whose name would be forever entwined with the story of the Donner Party.

Now, word of warning, if you want to view this document, don't come to the Library of Congress – because they don't actually have a copy and it's widely available online. I'm only here because I figured we needed a scene in a library to add intellectual weight to the content. So here I am.

SFX: leafing sheets of paper.

And here is the guide itself, which I printed off at home last night. It makes for fascinating reading, because it tells us so much about the man who wrote it. Of California – the state the Donner Party was headed to – he writes:

In my opinion, there is no country, in the known world . . . which is so eminently calculated . . . to promote the unbounded happiness and prosperity, of civilized and enlightened man.

But the real question is, why is Hastings so interested in selling California in such superlative terms. Does *he* have something to gain from the arrival of emigrants, and if so, what?

Here's Mrs Hoffman.

MRS HOFFMAN: We have to remember that in 1845 California is not a part of America. It's owned by Mexico. But Lansford Hastings – a lawyer from Ohio – is an ambitious man who has this dream of California becoming an independent republic, just like Texas was at the time. So a whole new country, in which he might hold the highest office.

Now in order to achieve that dream, first he needed more American settlers to reach his promised land to help overthrow the Mexican rule.

CALEB IN LIBRARY: This book that I'm holding in this sacred old library—

ANONYMOUS SHOUT IN BACKGROUND: Hey, buddy, keep it down!

CALEB IN LIBRARY:—was Hastings's attempt to lure people to California. But he also knew that one thing above all was going to put people off: the journey. The California trail was riskier and longer than the rival north-western route to Oregon. And so, in the final chapters of his book, Hastings came up with an ingenious solution – he proposed a shortcut. The only problem? He had never actually taken the route himself.

CALEB INTERVIEWING: So, talk me through the shortcut that Hastings proposed.

MRS HOFFMAN: Well, it looked good on paper, because—

CALEB VOICEOVER: She told me the route looked good on paper because Sutter's Fort – the place the Donners were heading in California, where Sacramento is today – was a straight line west of Independence, but the Oregon trail they'd be following for much of the journey took a big detour north-west, before diving back south.

Some of that detour was to get safely through the Rockies at South Pass, which was unavoidable, but even after that, the route continued even further north-west to Fort Hall, after which a fork in the road heading south-west from the Oregon Trail marked the beginning of the California Trail.

Again, you can check out the maps on my website to see what I mean.

MRS HOFFMAN: Hastings had—

CALEB VOICEOVER: Hastings had proposed cutting off this loop to save hundreds of miles of journeying. Viewed on a map, which in

1. Ikea's *Björksta* map (2019). 10/10 for colour, 0/10 for remembering New Zealand. *[See chapter 1]*

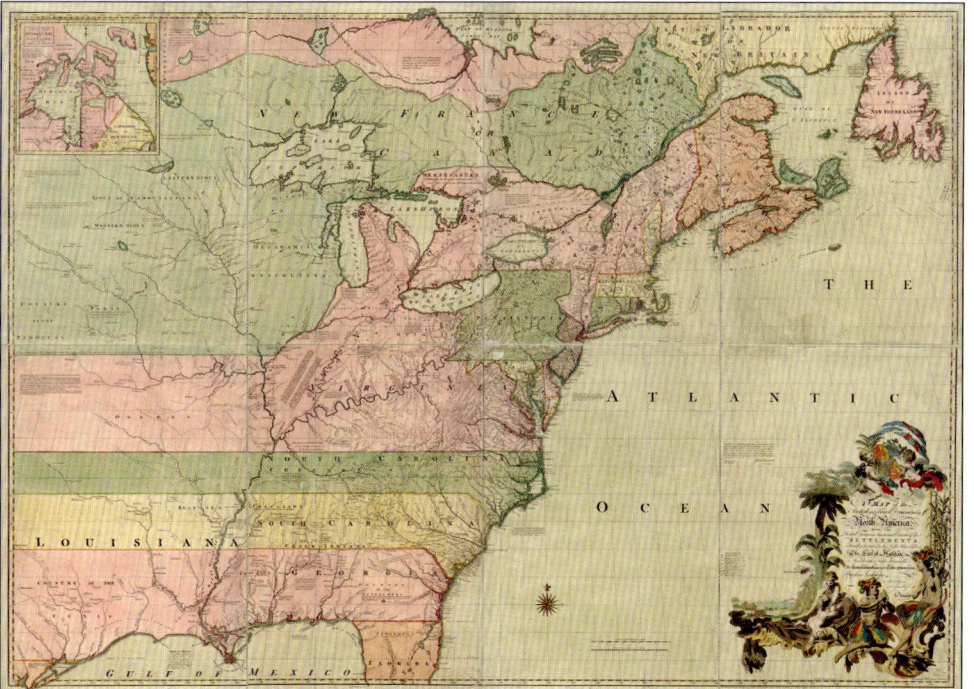

2. John Mitchell's map of the east of North America (1755). When western borders on the eastern States were a distant dream. *[Chapter 2]*

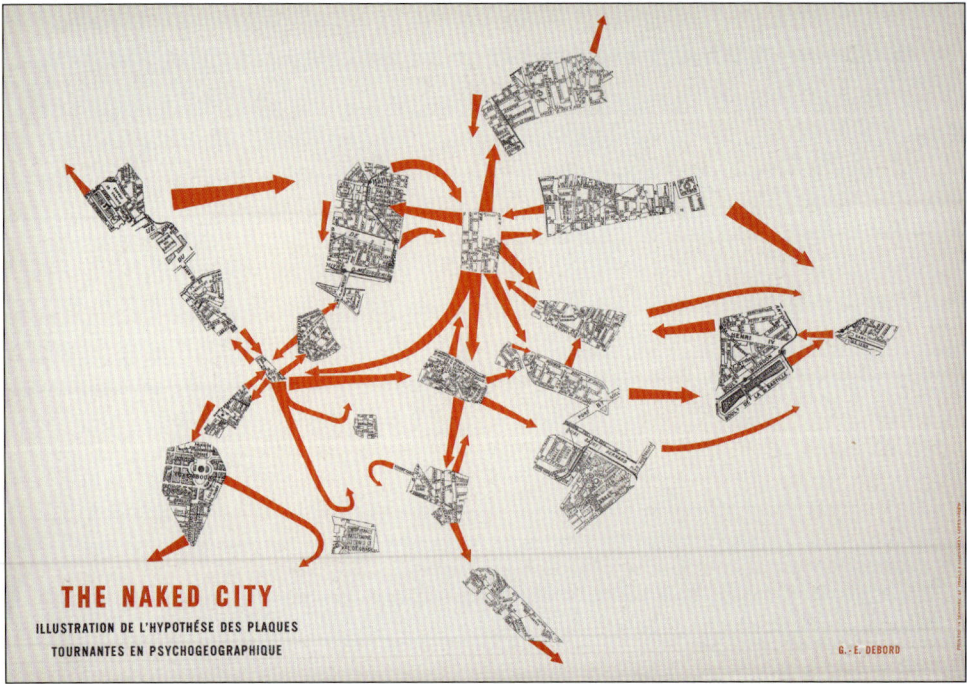

3. Guy Debord's *Naked City Map of Paris* (1957). Arrowy. *[Chapter 3]*

4. The 'Mountains of Kong' on John Cary's map of Africa (1805). *[Chapter 4]*

5. The 'Delta of Australia', according to Thomas Maslen (1833). *[Chapter 4]*

6. A very old map with Shetland in a box in (1736). Told you. *[Chapter 8]*

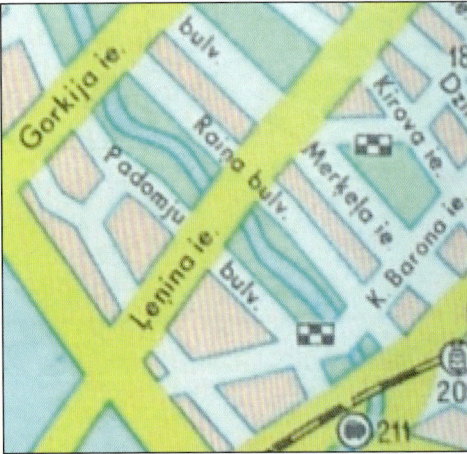

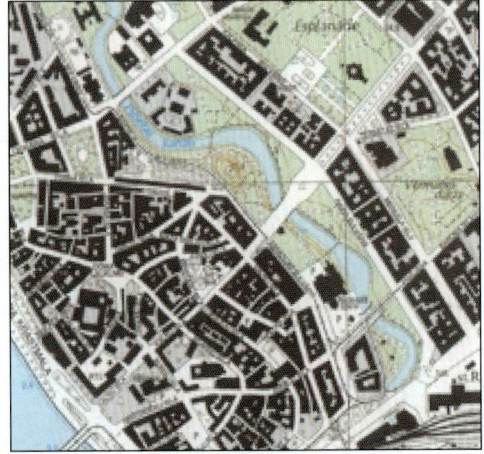

7. Two maps of Riga, Latvia, with very different levels of permitted detail. Left: on a Soviet tourist map (c.1980); right: on a post-independence topographic map (1991). *[Chapter 9]*

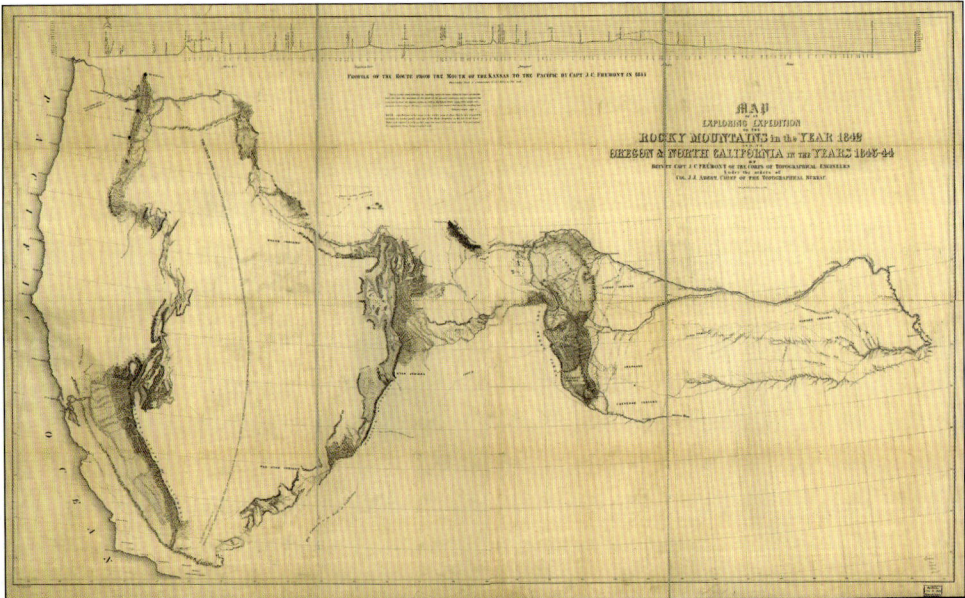

8. John C. Frémont's map of western America (1843). Lots of blank. *[Chapter 10]*

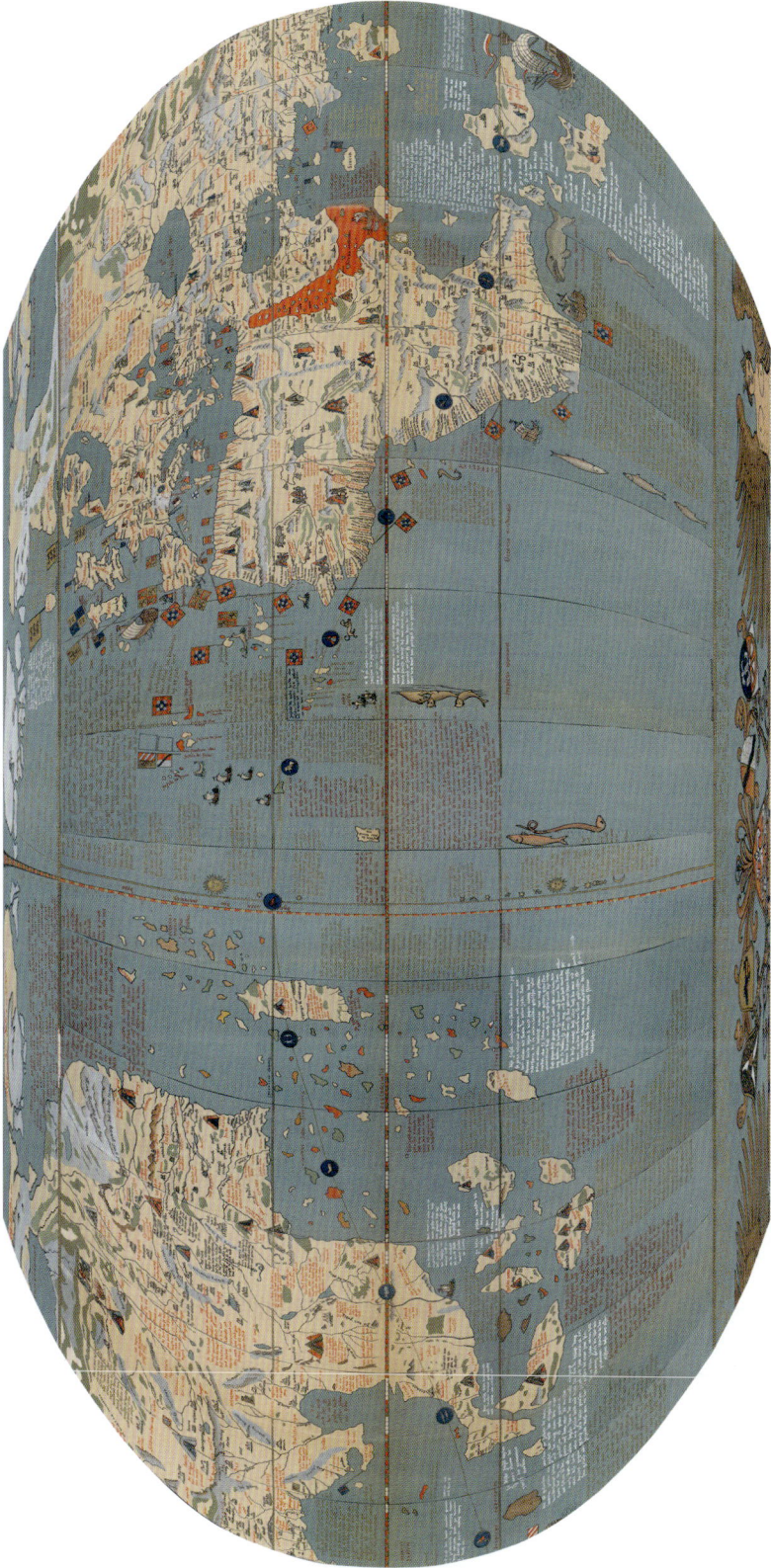

9. The 1492 *Erdapfel*, aka The Behaim Globe, but as a map (recreated 1908). *[Chapter 13]*

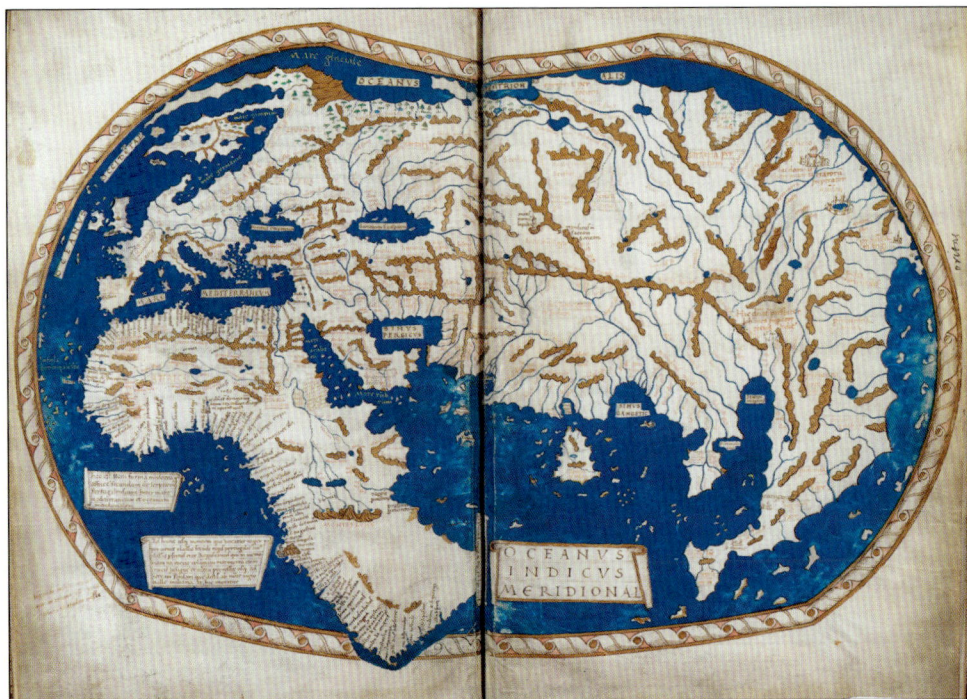

10. Martellus's world map (1489).
With Africa breaking the frame. *[Chapter 13]*

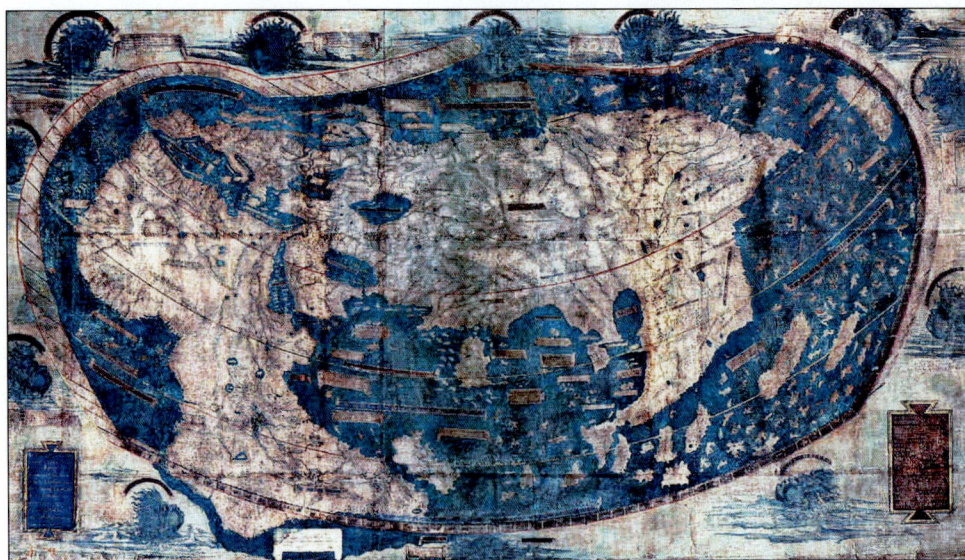

11. Martellus's other, slightly different world map (1490).
Worse condition but this one included the Atlantic Ocean.
The last surviving 'pre-Columbian' world map. *[Chapter 13]*

12. Martin Waldseemüller's world map (1507). The most expensive map ever sold, or, indeed, bought. Based on Martellus's earlier maps, Africa still breaks the frame. [*Chapter 13*]

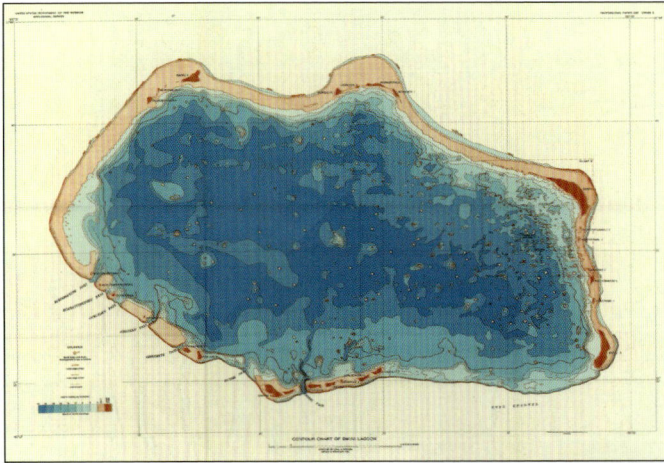

13. Map of Bikini Atoll, Marshall Islands, produced by the United States Geological Survey (1954). *[Chapter 14]*

14. Satellite image of the Castle Bravo thermonuclear blast crater on Bikini Atoll; still clearly visible (2017). *[Chapter 14]*

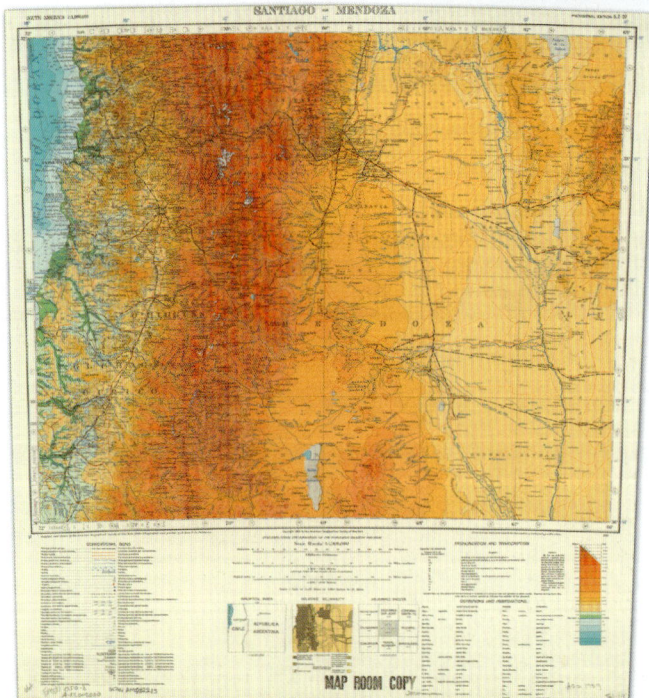

15. Sample sheet from the 'Millionth Map' project that was doomed never to be completed (1939). Shame, as it's rather pleasing to look at. *[Chapter 15]*

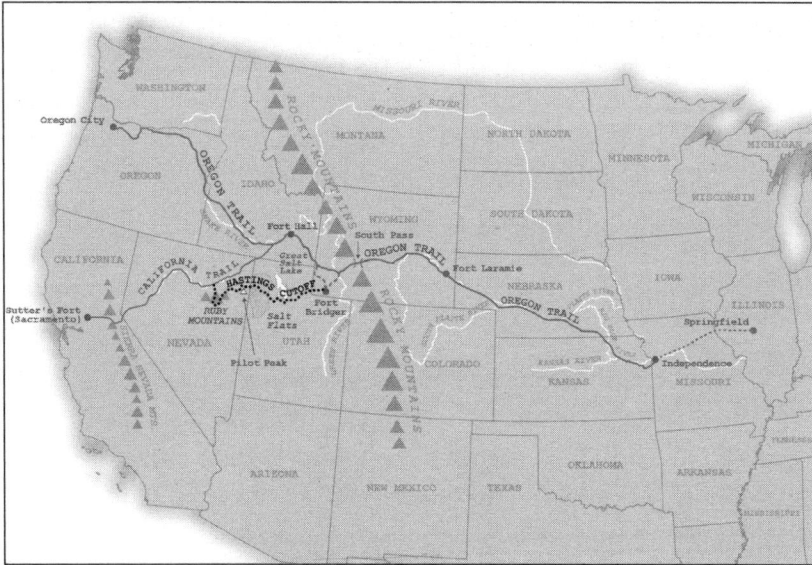

(*Courtesy calebcarterpresents.com.*)

those days wouldn't have shown any of the significant obstacles in their way, it would have looked pretty appealing. With one swish of his pen he'd made California seem way more inviting to potential emigrants who might previously have been heading to Oregon.

MRS HOFFMAN: And—

CALEB VOICEOVER: And even Hastings himself had no idea if it was a suitable route for those travelling with livestock and wagons, because when he wrote the *Emigrants' Guide* in 1845, he'd never set foot on it. In fact, hardly anyone had.

MRS HOFFMAN: If we look at the Frémont map, it clearly shows—

CALEB VOICEOVER: Annoyingly, Mrs Hoffman insisted on referring to the monkey map again, and was pointing to a big swooping arc of text written in a single curved line between the Rockies and the Sierra Nevadas, reading:

The Great Basin: *Surrounded by lofty mountains: contents almost all unknown, but believed to be filled with rivers and lakes, which have no communication with the sea, deserts and oases, which have never been explored, and savage tribes, which no traveller has seen or described.*

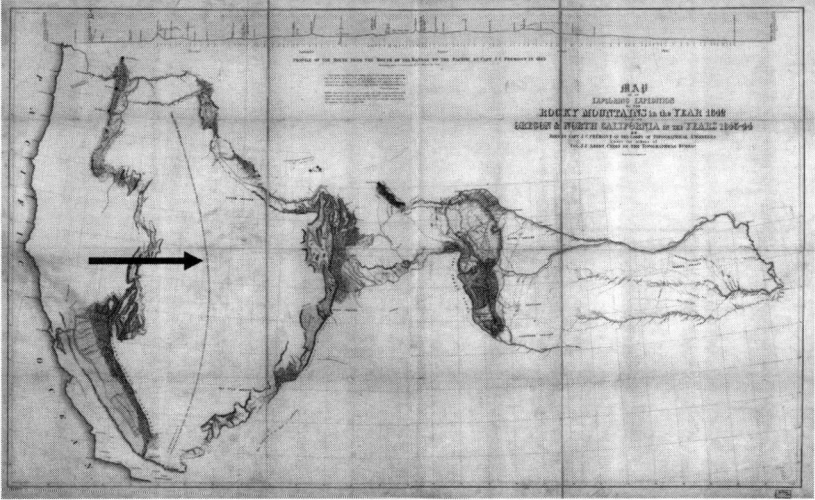

In short – no one had a clue what was there. As I sat in the library fingering the pages of Hastings's *Emigrants' Guide* – to create a sound effect – I couldn't help but think about the recklessness of what he was suggesting. Hastings was encouraging vulnerable families to risk their lives on a new path he himself couldn't vouch for.

It was only *after* the publication of this guide that Hastings first gave it a try.

SFX: horses' hooves on grassland.

In 1846, the year of the Donner Party journey, Hastings decided he would, finally, travel his own route, not once but twice. First, he'd travel east from California to Fort Bridger, on the far side of his shortcut – and in fact, as fate would have it, he set off east the very same week the Donners left Springfield, Illinois, in the opposite direction.

Then, at Fort Bridger, he hoped to personally convince as many emigrants as possible to return on the route with him, thus ensuring the arrival of more American-born families to aid his Californian Republic cause. As he set off from California, Hastings sent letters ahead to any oncoming parties announcing his promise to guide them on his new timesaving route.

But not all the information that disseminated from Hastings's journey east was to help his cause.

*　　*　　*　　*

CALEB VOICEOVER: After finally crossing the Big Blue River on their rafts, the Donners continued along the Oregon Trail, heading along the Platte River towards Fort Laramie on the edge of Wyoming.

And here, James Reed – the man I may or may not be related to – stumbled across an old friend, a man called James Clyman.

MRS HOFFMAN: And this is an incredible coincidence. Clyman had fought alongside Reed in the Black Hawk War many years earlier, and by pure chance had been travelling east from California *alongside* Lansford Hastings, so he himself had taken the Hastings Cutoff on horseback, the very first time it had ever been tested.

CALEB VOICEOVER: Clyman was a seasoned traveller and mountain man, the kind of guy you'd cast Jeff Bridges to play in a film. He basically *was* the landscape, only in human form. In short, if he had something to say about a journey across western America, you listened.

MRS HOFFMAN: And in no uncertain terms, Clyman warned Reed against taking the cutoff.

CALEB VOICEOVER: He tells him, 'Take the regular wagon track, and never leave it' (and yes, that's my best Jeff Bridges impression).

I have no idea what Reed thought after this meeting. Personally, I'd never ignore Jeff Bridges telling me anything. And I have to admit that this is the most frustrating thing about this podcast – I can't actually interview anyone about what they were thinking, or how they felt, because they're all dead.

I'd be lying if I said I didn't have a teeny bit of regret about not picking a contemporary story so I could interview some traumatised survivors or witnesses. I asked Ava if she had any bright ideas, and even though I think she only suggested a séance in jest, I told her she needed to grow up if she wanted a future in podcasting.

Whatever James Reed thought after hearing from Clyman, two weeks later the party encountered another eastbound rider, a loner with the super-quaint name of Wales Bonney. Wales was carrying one of Hastings's personal letters, about how he was going to guide anyone on his route himself from Fort Bridger.

It's easy to imagine how James Reed, the self-made mining man used to taking risks and finding reward, might have been swayed by this news. Likely he viewed the cutoff as a golden chance to make up lost time.

MRS HOFFMAN: At this point, decision time is coming up fast. On July 18 the Donners and Reeds crossed the Continental Divide in the Rocky Mountains, a hugely symbolic moment, because they were now in the part of America where water flows into the Pacific Ocean instead of the Atlantic.

CALEB INTERVIEWING: Excuse me?

CALEB VOICEOVER: Mrs Hoffman explained that the Continental Divide, which I'd always assumed was a band my dad listened to, was the natural line running down America – all water on one side of it flowed east; on the other, it all flowed west. This was nothing short of incredible to me – like, how can water be so *decisive*?

Anyway, this meant the Donners were now walking with the flow of the rivers, but also . . .

MRS HOFFMAN: Having crossed the bottleneck of the South Pass, they faced a choice: carry on the traditional, trusted path that went north to Fort Hall then on to the California Trail turnoff, or bear southwest now towards Fort Bridger, and Hastings's new cutoff.

CALEB VOICEOVER: Some were sceptical. Perhaps they looked at a map like Frémont's, and baulked at the vast unknown they'd be stepping into. But in the end James Reed's stubborn determination to gamble won out. George Donner gave the go-ahead and his eponymous party?... turned left.

*   *   *   *

    SFX: phone rings.

CALEB ON PHONE: Hello?

FEMALE CALLER: Hello, is this Mr Carter?

CALEB ON PHONE: Yeah, look, I thought I told you to delete my number.

FEMALE CALLER: This is ancestry.com.

CALEB ON PHONE: Oh . . .

FEMALE CALLER: We've got your results . . . we think we've found something you might be interested in.

    Caleb gasps.

CALEB VOICEOVER: This time, I *was* shocked. I *didn't* know this was going to happen. It had been so long since I'd sent them my DNA sample, I simply hadn't expected to hear back. And now? I just wanted to hear what they had to say. Might I be distantly related to the Reed family, creating such a personal connection to this story it would all but ensure critical success for the podcast?

FEMALE CALLER: First off, Mr Carter, we wanted to mention your sample. We just needed a swab, OK? Four cups of saliva swashing around inside a Wendy's burger box wrapped in cling film? That was gross. I have no idea how long it took you to do that, or why.

CALEB ON PHONE: It took me two hours. I just wanted to be really sure you had *all* my DNA.

FEMALE CALLER: That's not how it works.

CALEB ON PHONE: You said you had something interesting?

FEMALE CALLER: Yeah. So based on our records we've found a number of your descendants, going back decades . . .

CALEB VOICEOVER: My heart was pounding, I just wanted to *know*. This was the moment that would potentially redefine not just the success of *Left, for Dead*, but my entire sense of self.

FEMALE CALLER: And when it comes to characters from the story you're researching, we can say with almost 100 per cent confidence that your great, great, great, great grandfather was the cousin of the wife of the brother of Lansford Hastings.

    Brief silence.

CALEB ON PHONE: What was that?

FEMALE CALLER: Lansford Hastings? You're a direct descendent of his brother's wife's cousin.

CALEB VOICEOVER: For probably the first time in my life I was speechless. I didn't know how to feel. On the one hand, it was undeniably great that I did have a personal bloodline connection to this incredible story. And such a close one at that. On the other, Lansford Hastings was quite clearly the villain, the man whose own greed for power led to unspeakable tragedy.

I knew this revelation would take some time to process. Most likely the exact amount of time between recording Episodes 1 and 2. Right now, I just didn't know what to say. So I said nothing.

FEMALE CALLER: Oh, and sir? Your credit card payment didn't go thro—

CALEB VOICEOVER: And then I hung up.

*   *   *   *

CALEB VOICEOVER: The Donner Party had turned left, and were now marching towards Fort Bridger, where they would meet Lansford Hastings. But they hadn't yet passed the point of no return, because if they decided they didn't like him, they could still head back north from Fort Bridger to meet the main trail again. Again, you can check out the maps on my website to see what I mean.

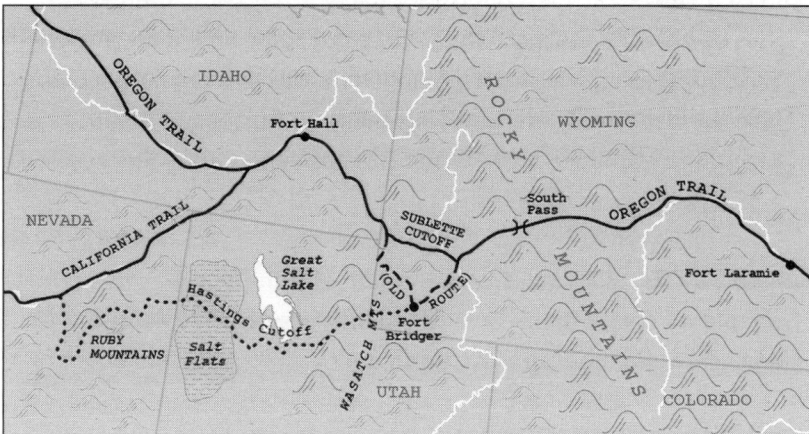

(*Courtesy www.calebcarterpresents.com.*)

But by the time the Donner Party reached Fort Bridger, Hastings had already gone – setting off to guide the Harlan–Young party, who had also gambled on the cut-off, and who (like everyone) were ahead of the Donners.

Hastings's absence must have been a huge blow to James Reed. And now, at the fort, they faced an even bigger decision. Stick or twist? Return north to the tried-and-tested path, or take the shortcut alone and hope for the best?

MRS HOFFMAN: We know that Tamsen Donner, party leader George Donner's wife, was strongly in favour of the conventional path. This was a woman with three young children, who had previously lost a husband and child many years before her marriage to George. She had taken years to rebuild her life to this point and had, perhaps understandably, been hesitant about the migration from the start. Now, Tamsen was determined to protect her young children against this dangerous-sounding suggestion.

CALEB VOICEOVER: Perhaps predictably, the proud James Reed still insisted they should try to make up time. The group was divided. But among the different opinions, fate was to play a final role.

MRS HOFFMAN: So, there's this newspaper man called Edwin Bryant, who had been travelling with the party for some time, only to grow frustrated with the slow pace. Unencumbered by a family of his own, he'd decided to split off, travelling ahead via horseback. Bryant reaches Fort Bridger ahead of the Donners, and speaks there with a man called Joseph Walker.

CALEB VOICEOVER: Walker was another Jeff Bridges type – lots of people were back then – and, like Clyman, he strongly warned against taking the Hastings Cutoff.

MRS HOFFMAN: Bryant, now concerned about Reed's headstrong plans, took it upon himself to write his former travelling companions a letter, dissuading them from continuing with the shortcut, and urging them to head back to the nearby Oregon trail to the north.

He left his letter with Jim Bridger, who owned the eponymous fort, and rode on.

But Bridger, knowing the contents of the letter, faced a dilemma of his own. His fort had been built on the main trail just three years earlier, and selling goods to weary travellers was key to his livelihood. Unfortunately for him, just a year later the common path had changed course, as a new shortcut called the Sublette Cutoff offered a drier, dustier and two-and-a-half-days-quicker route bypassing the fort, and cutting off his source of income.

The new Hastings's cutoff, should it prove a success, would put Fort Bridger back on a recognised path. Bridger would have been desperate for it to succeed.

With this no doubt forefront in his mind, Bridger ensured the letter never made it into the hands of the Donner Party.

Without its crucial information, Reed's bullish conviction won out once again.

'And besides, anything to avoid Idaho', he may have remarked, as the clincher.

The next morning, the Donner Party rounded up the livestock, hitched their wagons, and set off on what were to be the most gruelling, energy-sapping and torturous miles of the entire journey.

They had decided to brave the unknowns of the map, and risk the Hastings Cutoff.

SFX: thunder claps. Theme music rises.

*   *   *   *

CALEB VOICEOVER: In the final episode of *Left, for Dead*, the Donner Party come to terms with their decision, as they are confronted with some of the perils of the American wilderness; I have an epiphany about Lansford Hastings; and the podcast takes a step closer to awards glory as things start to get spooky.

CALEB OUTDOORS: I swear to God I'm not moving the counter, it's doing it *by itself.*

CALEB VOICEOVER: Till next time.

*   *   *   *

CALEB VOICEOVER: Have you ever found yourself in a handshake that goes on a bit too long, but you're not sure how to politely release yourself from your acquaintance's vice-like grip? Then you should try Hank's Handy Handshake Oil, a slick and shiny ointment that will help you free your hand from any handshake, no matter how determined your adversary is to keep going. With Hank's Handy Handshake Oil you can slip out of their muscley finger-hold as soon as you're ready. Simply spray it onto your right hand first thing in the morning, or in the waiting room right before that job interview. With an oily palm, you'll never feel awkward again.

Subscribe today and get 20 per cent off with the promo code OILYCALEB.

Hank's Handy Handshake Oil: handing you control.

*   *   *   *

Episode 2

A rustle of wind in the microphones, the swish of long grass and chirping of crickets. The occasional car and motorbike go past. Footsteps on tarmac.

CALEB OUTDOORS: Is that it . . .? It must be. I think that's it.

Pause.

Doesn't look like much, does it?

AVA OUTDOORS: Sure is run down.

CALEB OUTDOORS: It's kind of amazing, though.

CALEB VOICEOVER: I was standing with my producer, Ava, on County Road 493 near the mouth of the Sacramento River in California, looking across at a patch of private land beyond a quaint rural fence.

CALEB OUTDOORS: It's a shame we can't get any closer, I guess.

AVA OUTDOORS: Yeah . . . but wait – why are we here?

CALEB VOICEOVER: And I was transfixed, because between a thicket of trees I could just about catch glimpses of an old building. A flank of corrugated roof; a patch of faded white paint on weatherboard. And while I couldn't make out much, I could tell it was pretty run down.

CALEB OUTDOORS: It's a shame nobody's looking after it more.

AVA OUTDOORS: I guess no one really cares.

CALEB OUTDOORS: Yah. So sad.

CALEB VOICEOVER: This dilapidated building was once known as 'Hastings Adobe', the ranch that belonged to Lansford Hastings when he moved to California. With no known burial site, it was the only place I could think of visiting to commemorate Hastings. The man whose brother's wife's cousin was my great, great, great, great, grandfather. A man I increasingly felt had been treated incredibly unfairly by history. Including the history of my podcast – Episode 1.

AVA OUTDOORS: So, have we got what we need?

CALEB OUTDOORS: I feel like I'm 'home', ya' know?

CALEB VOICEOVER: Sure, old LH is remembered for selfishly leading dozens of innocent people through an impossible trail against the advice of everyone he spoke to, at the whim of his own political ambition, resulting in the greatest human tragedy in the history of US emigration. But at the same time, *I* was *related* to him.
   And that had to count for something.

CALEB OUTDOORS: I just wonder if there's a different truth about Lansford Hastings out there, and we're the ones to find it.

AVA OUTDOORS: Interesting. I guess, having done a lot of reading, I don't think—

CALEB OUTDOORS: I've got an idea. Come with me.

AVA OUTDOORS: Where are we going?

   MUSIC: theme creeps in, gradually building.

CALEB VOICEOVER: All of a sudden, I knew what I had to do.

CALEB OUTDOORS: We're crossing the Sierra Nevadas.

AVA OUTDOORS: Uhhh . . .

CALEB VOICEOVER: It was time for me to immerse myself in the story of which I was now so strongly a part, by crossing the mountains that bordered California – by car, of course, I loathe hiking – to a lake once known as Truckee Lake. But today? It goes by the name of Donner Lake.

Maybe there I'd find some clues and finally reveal the truth about my long-misunderstood relative, the great explorer and trail-finder Lansford Hastings.

MUSIC: theme reaches a climax, and abruptly cuts.

CALEB VOICEOVER: Episode 2. Party's over.

*   *   *   *

CALEB VOICEOVER: The Donner Party had – not exactly collectively – decided to try the new shortcut proposed by Lansford Hastings instead of sticking to the traditional route that looped up to the north.

And now, with no map to guide them, they were relying on a mixture of the tracks in front of them for where to head.

Here's Mrs Hoffman, my old social studies teacher, picking up the story.

MRS HOFFMAN: The first hurdle on the Hastings Cutoff was the Wasatch Mountains, but the party Hastings was guiding up ahead had taken a difficult route down a steep canyon. So, Hastings – aware there's another party following behind – writes a letter and ties it to a bush, telling them not to follow. And the Donners somehow manage to find this letter.

CALEB INTERVIEWING: Oh really? So . . . what – Hastings was just really good at leaving letters in clever places?

MRS HOFFMAN: Well, it was just chance that they—

CALEB INTERVIEWING: He must have picked a great spot, though, and tied it super-tight. Sounds smart. Go on.

MRS HOFFMAN (inhales deeply): Anyway, Hastings suggested in his letter they send a rider ahead to find him, promising he'd return to guide them along a better route through the Wasatch Range.

CALEB INTERVIEWING: That's very kind, don't you think?

MRS HOFFMAN: *But* – when James Reed rides on and finds him, Hastings goes back on his word and refuses to return to the party, instead pointing out a rough direction where they should try to cross, from afar.

CALEB INTERVIEWING: I guess that's all they needed, though, right? Plus, he had to look after the Harlan–Young party, who were *not* running irresponsibly late.

MRS HOFFMAN: Don't argue with me, Carter. Hastings is repeatedly breaking promises—

CALEB INTERVIEWING: *Or,* he's responsibly reassessing the situation at every moment.

MRS HOFFMAN: And you've got a PhD in this now, do you? You know, it was naïve of me to think you might have matured a single day in 20 years.

CALEB INTERVIEWING: I'm only interrogating the evidence, like you always taught.

MRS HOFFMAN: Don't be such a creep. You know what—?

Muffled sounds of a radio mic being ripped off.

I'm done. Enjoy researching the story alone, Carter – and yeah, that'll mean actually reading some source material. Good luck with that.

Sound of footsteps and a door slamming.

⋆   ⋆   ⋆   ⋆

CALEB VOICEOVER: For the second time in my life, Mrs Hoffman had walked out on me. And there was no denying that this was a blow – she knew a lot about the subject – even if she was being kinda obtuse.

I didn't feel she was being fair to Hastings. And now, armed with just Ava's summary notes, I had a job to do on my own.

When Reed returned to the party after the only encounter any of them would ever have with Hastings, he did his best to guide them through the mountains as instructed. But it's fair to say, this leg of the journey was nothing short of a total disaster, 'cos it turned out the route Hastings suggested? It was pretty impassable. Then again, maybe Reed just wasn't concentrating hard enough when Hastings pointed out the best route. We can't count that out.

As they attempted the crossing, steep inclines and descents made the going almost impossible for the livestock hauling the wagons. Worse still, the trees and undergrowth got so thick that they sometimes had to stop for days at a time just to cut down trees and clear away bush to make a path. It was gruelling work, where for a time they only made a mile and a half's progress a day.

And when they finally did make it across the Wasatch? An even greater challenge awaited: the vast salt plains of northern Utah. A lifeless desert that included a long, perilous stretch without a drop of water, where struggling animals often sank into the slimy alkaline bog lying just beneath the baked surface of the flats.

This is where the Reeds found themselves alone and stranded, desperately awaiting rescue, but of course all the families suffered on this leg. When they finally made it to a spring beside Pilot Peak on the other side, 36 cattle had been lost, with a number of wagons – containing vital food and belongings – left abandoned in the desert.

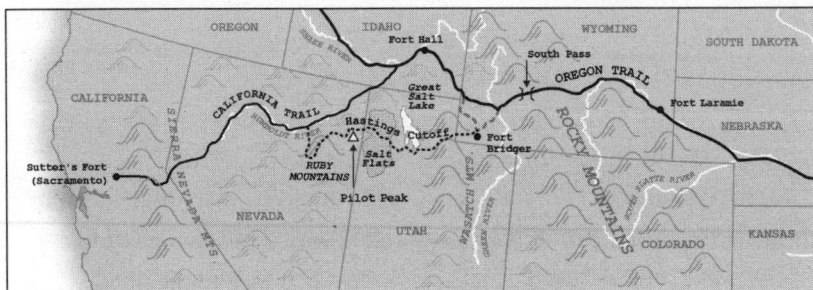

(Courtesy www.calebcarterpresents.com.)

Years later, George Donner's youngest child Eliza would write of this ordeal:

> Anguish and dismay now filled all hearts. Husbands bowed their heads, appalled at the situation of their families. Some cursed Hastings [. . .] Mothers in tearless agony clasped their children to their bosoms, with the old, old cry, 'Father, Thy will, not mine, be done.'

And yet, there was still one final sting in the tail of the Hastings Cutoff. After days of trudging through the eastern hills of Nevada, they soon squared up to the Ruby Mountains. Hastings, up ahead, had no idea that the best way through was nearby to the north. Instead, he guessed on turning south with the Harlan–Young Party, forcing them to march alongside the mountains for three days until they found a suitable right hand turn to navigate through. And the Donners? They had little option but to follow in their tracks, wasting yet more valuable time.

When leaving Fort Bridger before the cutoff, Reed had guessed they'd be in California in seven weeks. Two months later, the Donner

Party rejoined the California Trail at the Humboldt River, physically and mentally broken, yet with hundreds of miles still ahead of them. Everyone now knew that, having lost so much time, they had to move fast if they were to cross the Sierra Nevadas before the deep snows arrived.

*   *   *   *

SFX: walking boots on a worn footpath.

CALEB OUTDOORS: So this is it – Donner Lake.

SFX: backpacks and camping equipment falling to the floor.

AVA: So, shall we just set up camp here?

CALEB OUTDOORS: Yeah, yeah – go ahead. I'm just gonna sit down and breathe it in.

CALEB VOICEOVER: Ava and I had trekked to Donner Lake from Donner car park, a few dozen yards to the west. Everything about this place was like a memorial to what happened here – there was Donner Trailhead, Donner Woods, Donner Lake Boat Launch, Donner Ski Ranch. I felt sad the Donners never got to see all this amazing stuff honoring the spot where loads of them died and ate each other.

When the Donners reached this spot on October 31, 1846, the party had 81 members. Over half were under 18, and a quarter were younger than five, including six babies.

Eesh.

Since completing the Hastings Cutoff and reaching the lake via the remaining section of the traditional California Trail, there had been no let-up in their bad luck. Raids by Native Americans along the Humboldt River saw valuable livestock stolen, and in a bizarre moment of madness James Reed murdered another family's teamster after a heated argument, resulting in the banishment of the party's most natural – albeit contentious – leader.

Now, at the end of October, there was just one final piece of American geography to conquer: the Sierra Nevadas. A hundred and eighty years after the Donners reached them, I was looking up at the same awe-inspiring mountains from the very same spot.

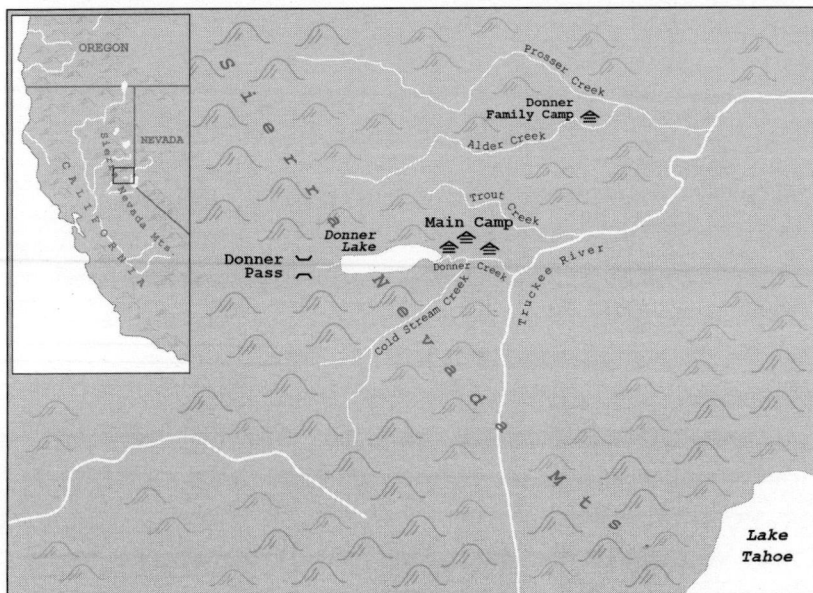

(*Courtesy www.calebcarterpresents.com.*)

SFX: deliberate trudging of hiking boots.

CALEB OUTDOORS: I'm just below Donner Pass, the 7,000ft elevation the party had to climb to break the back of the Sierra Nevadas. And, as they arrived on October 31, conditions were good to do so.

But, weary from the day's march, they made one final, fateful decision. Instead of ploughing on up and setting up camp on the other side, they stayed here by the lake, too exhausted to climb that evening.

Had they known the English translation of Sierra Nevada – 'Mountains Snowy' – they might not have hedged their bets. Because that night, a blizzard laid a thick blanket of snow on the pass, making it impassable.

By not crossing when they had the chance, by not arriving just a half-day sooner, the Donner Party had missed the season's last window to make it over the mountains. Because now, an early and bitter winter had arrived.

*   *   *   *

MUSIC: a minor-key variation on the podcast theme music, suggestive of impending tragedy.

CALEB VOICEOVER: The men now set about building temporary cabins down by the lake, some of which were shared between families. The Donner family were a few miles further back at Alder Creek, after an axle had broken on one of their wagons. In the process of fixing it, George's son had swung an axe into his father's wrist, creating a dangerous open wound that would soon become infected.

MRS HOFFMAN: And, of course, the biggest challenge by far would be food.

CALEB INTERVIEWING: Mrs Hoffman, good to have you back.

MRS HOFFMAN: Yes. Well, I thought about it and realised that . . . perhaps you had a point, Carter. Perhaps I was unfair on you. Also, professionally, I felt it was important I see this thing through.

CALEB VOICEOVER: Ava said that for ethical reasons, I should declare that Mrs Hoffman had not actually come back in her entirety. Instead, I used our previous interviews to recreate her voice using AI.

I'd come to realise I still needed her expert voice in the podcast, but rather than having to find ways to edit her words to suit my narrative, it was way easier just to have Mrs Hoffman say exactly what I wanted her to say by just typing it in to a computer.

MRS HOFFMAN (AI): Sorry for being confrontational earlier. I've come to realise I've never given you due credit, not in school, and not now. I'm sorry.

CALEB INTERVIEWING: That's kind, Mrs Hoffman, but let's just get back to the story.

MRS HOFFMAN (AI): Sure. So, in their camp towards the foot of the mountains, the party were already dangerously low on supplies, the livestock emaciated by the brutal journey. Some families had handfuls of flour, but one simple fact was unavoidable: there wasn't enough food to go round. And, after a particularly long and brutal blizzard towards the end of November, there was even less, because many of the remaining animals got buried in the deep snow.

CALEB VOICEOVER: This was so cruel. The animals were not only dead, but the snow was so deep that no one could find them to eat them, despite their best efforts poking around with a long stick.

A few weeks into the ordeal, even a simple search for firewood around the cabins became impossibly energy-sapping. Many a night was spent without a fire, with young children on empty stomachs wrapped in frozen wet blankets.

Myself, I happen to know a thing or two about how they must have felt, longing for proper sustenance. A couple of years ago I did a 5:2 diet, where two days a week I'd eat fewer than 500 calories in a day. On those excruciating days I'd be limited to black coffee, a few nuts, some carrot sticks, and a couple of poached eggs with zucchini in the evening. Maybe a hollandaise on a cheat day.

I swear to God, I nearly lost my mind. I was getting headaches, seeing double and my pee tasted super-weird. Still, I lost four pounds in a month, and almost managed to get back with my ex.

For the members of the Donner Party, huddled around their occasional fires in makeshift cabins, too weak to venture outside to use the bathroom, it was probably even worse.

And then, the news everyone was dreading but knew was coming.

*   *   *   *

SFX: a busy kitchen, frying, burgers being flipped, saucepans on stoves and crockery being stacked.

CALEB IN DINER: Yeah, medium rare, thanks. And a Diet Coke, please.

I'm at the Donner Lake Kitchen, something the Donner Party must have wished was here two centuries ago, but would have been weird if it had been – especially with that name.

Because by December 15 a young man named Baylis Williams – who worked for the Reed family – passed away, most likely from starvation. His death prompted something extraordinary. Because the next morning, on the 16th, a party of the fittest men were spurred into action. They decided they simply had to brave the mountain crossing and try to rally help on the other side before they all met the same fate as Baylis.

The group of 17, who became known as the Forlorn Hope, fashioned snowshoes to help them ascend, and although two of them turned back early on, the rest made good progress.

Five days in, however? Fatigue and hunger were taking their toll. One member of the group was left behind, too weak to continue, then on Christmas Day two others died when they were trapped in another blizzard.

The remaining members of the Forlorn Hope were now presented with a conundrum: risk starving to death beside their companions . . .

Or eat them?

SFX: plate being put down.

WAITRESS: There you go.

CALEB IN DINER: Smells great, thank you.

    SFX: cutlery on plate.

CALEB IN DINER: Perhaps they thought long and hard about it, perhaps they barely hesitated. But in the end they did what they had to do. The first members of the Donner Party tasted human flesh. And they wouldn't be the last.

WAITRESS: Everything OK with your steak, sir?

CALEB IN DINER (mouth full): Delicious, thanks so much.

CALEB VOICEOVER: On January 17, a month after the Forlorn Hope had set off from Donner Lake, one member of the escape group, William Eddy – having walked 18 miles on bloodied feet that day – reached a settlement called Johnson's Ranch. Some way behind him were a further seven survivors, whom the inhabitants of the ranch rode out to rescue.

Seven more had perished on the journey. Because it turns out that sometimes even eating your friends isn't enough to see you through.

       *   *   *   *

CALEB VOICEOVER: Meanwhile, on the eastern flank of the mountains the situation was becoming increasingly desperate.

MRS HOFFMAN (AI): Families had now resorted to boiling down the hides of oxen and mules into a gluey paste and spooning it to their young, with everyone doing their best to hold down the thick, grey mixture.

CALEB VOICEOVER: More people died, and bodies soon piled up outside the cabins. You've probably guessed what happened next, because the temptation to use those bodies for nourishment became overwhelming.

It's funny, but the words 'they ate each other' feel so insufficient for capturing the horror, shame and moral complexity – perhaps even relief – of that decision. Once again, I wished I could interview someone to help me understand this pivotal moment.

Then all of a sudden, it hit me.

*   *   *   *

SFX: the urgent tread of hiking shoes.

CALEB OUTDOORS (shouting): Ava!

AVA (from afar): Yeah?

CALEB OUTDOORS: I've had an idea, a way for our podcast to say something new about the story. Get the candles.

*   *   *   *

SFX: tent being zipped up from inside. Wind howling.

CALEB IN TENT: Okay. Ready?

AVA: You know, I *was* only joking when I suggested this.

CALEB IN TENT: Turn the mic sensitivity to max – we need to capture every sound.

CALEB VOICEOVER: In my hand was a smooth, grey stone I'd picked up from beside the lake, which I now placed at the centre of my homemade Ouija board. I knew if I could find a way through the liminal space separating this life from the next, I had a chance of scoring an exclusive interview with a member of the Donner Party. And who better to approach than the party leader himself?

CALEB IN TENT (with a soft, somewhat am-dram intensity): O Spirits of the Lake, we welcome you into our hearts, our minds and our tent. We call on George Donner to move among us and speak through us. Oh my God, did you see that?

AVA: See what?

CALEB VOICEOVER: The candle flickered powerfully.

CALEB IN TENT: It's George. I can feel him here. George? Look, I'm so sorry about what happened, man.

     Slight pause.

AVA: OK, well – I guess we just go back to my suggested script I sent?

CALEB IN TENT: Oh. My. God!

CALEB VOICEOVER: A shiver went up my spine as the stone started to move. This was my first communication with the dead, not to mention the first ever interview with George Donner.
    His first words? Full of trademark gratitude and simplicity, they spelt out:
    THANKS MAN
    George and I began to shoot the breeze, building an instant and undeniable rapport before I got down to business.

CALEB IN TENT: George, I gotta ask you – the decision to eat other party members? What did it take to cross that line?

CALEB VOICEOVER: There was a considered pause, and then?

CALEB IN TENT: OK, it's moving again.

CALEB VOICEOVER: George's answer was something that would literally never occur to me. He said:
OF COURSE, I NEVER WANTED TO SWALLOW MY DEAR DEAD FRIENDS, BUT THEN I ASKED MYSELF: 'IF I WERE DEAD, WOULD I RATHER MY BODY BE PART OF SOMETHING DEAD, OR PART OF SOMETHING LIVING?' PSYCHOLOGICALLY, THAT HELPED.
PLUS, I WAS REALLY, REALLY HUNGRY.'
And that was it.

CALEB IN TENT: Wow.

AVA: Yeah, interesting.

CALEB IN TENT: I know, right?

AVA: You know, from what I read there's no evidence George Donner actually ate anyone.

Pause.

CALEB IN TENT: Yah ... huge exclusive.

*   *   *   *

CALEB VOICEOVER: And speaking of exclusive, Left, for Dead has a new exclusive deal for all our listeners from our sponsor.
We all know stress can be hard to manage. But whenever stress starts getting on top of me, I use Hank's Handy Stress Syrup. Just one spoon of Hank's Handy Stress Syrup and you'll be out cold for up to 36 hours. By the time you wake up you'll barely remember your own name, let alone any of your problems. It's the perfect way to press the 'off' switch on life, and forget memories.

Sign up now with the promo code BRAVENEWWORLD for twice-daily marketing emails.

Hank's Handy Stress Syrup: Leave life behind.

*   *   *   *

CALEB VOICEOVER: The desperate man-eating families by Donner Lake had no idea that the Forlorn Hope had, by the skin of their teeth – not to mention the skin of other party members – made it to civilisation, from where they could alert people to the party's fate and mount a series of rescue missions.

MRS HOFFMAN (AI): But, of course, a rescue was no simple thing. The task of guiding weakened bodies across the perilous mountains meant that relief parties faced enormous risks themselves.

CALEB VOICEOVER: But William Eddy and John Sutter, owner of Sutter's Fort in modern-day Sacramento, managed to rally some men to the cause, as did the banished murderer James Reed, who'd heard of his family's desperate situation and presumably wouldn't be so banished if he now returned to save them.

MRS HOFFMAN (AI): When the first relief party made it down to the camp by the lake they were met with a hellish scene. Sordid hovels inhabited by emaciated ghosts were surrounded by bodies, many of whom had clearly been cannibalised.

CALEB VOICEOVER: And the families faced impossible questions: who went back first, and how many? Did mothers brave the journey with infants? Or did they send their young children with the strangers, possibly never to see them again?

The stories of the many rescue efforts that relayed across the Sierra Nevadas could fill an entire podcast on their own, including the group of 13 mostly women and children who became stranded in the mountains by a storm, with just a fire and no shelter for almost a week. Miraculously, 11 of them made it down the other side.

But amid the tragedy there were heartwarming moments, such as when 13-year-old Virginia Reed and her mother Margret, who left with the first relief party, staggered down the western slopes of the Sierra Nevadas in mid-February only to run into James Reed, whom they hadn't seen for months, himself heading over on his first rescue mission. The long embrace between the young Virginia and her stepfather would have been enough to thaw any frostbitten extremity.

But Margret had been forced to leave two of her children – aged eight and three – behind, so despite the emotional reunion James continued up the Sierras, ultimately returning as close to death as any man, but with both children alive.

And the Donners?

MRS HOFFMAN (AI): George's wife Tamsen actually refused rescue, staying with her husband in their makeshift camp a few miles back from the rest. George's infection from the axe wound left him too ill to move, and for Tamsen, perhaps the idea of losing two husbands in one lifetime was too much to bear. Although the couple didn't survive, all five of her children made it to California.

CALEB VOICEOVER: In April 1847 the last survivor of the party was rescued, a German named Lewis Keseberg. Of the 81 who had arrived with the Donner Party at what was then Truckee Lake, only 45 made it to California alive.

Just over a year after the Donners had set off from Springfield, Illinois, their journey was finally over.

*   *   *   *

SFX: clicking of mouse and typing.

CALEB VOICEOVER: I'd been thinking a lot about what the real Mrs Hoffman had said – about researching the story myself and looking at source material. And to be honest? The thought made me feel physically sick.

At the same time, having reached the end of the story, I realised it still sounded annoyingly like Hastings had been kind of to blame. His route had undeniably added time to their journey. And that time? It had proved pretty critical.

But how could I be sure I hadn't missed something? That Ava's summary notes I'd been relying on to tell the story without Mrs Hoffman might not have overlooked something? Maybe me actually doing some work was all that stood between the tidal wave of historical consensus and an incredible exoneration.

I decided to start by opening the 1845 map Ava had sent me near the start of my journey – the one based on the expeditions of John C. Frémont and which looked like a monkey – with its blank space into which Hastings had blindly dared to venture.

SFX: typing.

CALEB VOICEOVER: But when I typed in the words 'Frémont' and 'map', I found something I wasn't expecting.

CALEB IN OFFICE: What the hell? 'Map of Oregon and Upper California – 1848?'

CALEB VOICEOVER: In front of me was another map of Frémont's expeditions, way more detailed than the monkey one, with America way more recognisable. Check it out on the website.

It turned out that Frémont loved exploring this part of America and had done loads more of it since he made the first map, published in 1845.

It felt like this could be important, but I had no idea how. I tried to concentrate.

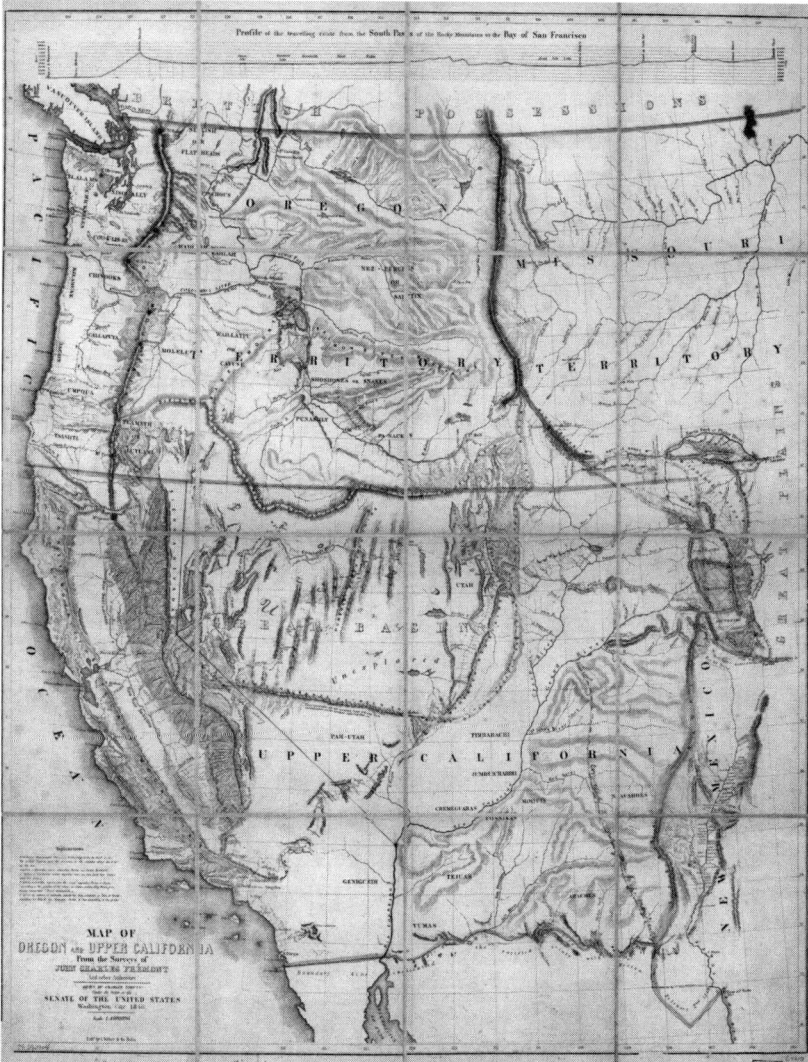

(Courtesy www.calebcarterpresents.com.)

SFX: sound of coffee cup being slammed.

CALEB VOICEOVER: I drank a quadruple espresso, and after an intense migraine and some stomach cramps, I actually did concentrate. Hard. And after two whole days of staring at the map without eating or sleeping, I had a breakthrough.

CALEB IN OFFICE: So . . . there are these faint dotted lines on the map which, I *think*, from the key, are the different routes taken by Frémont on different expeditions. And what's interesting is there's one here right across the salt flats to the springs at Pilot Peak, going through the Ruby Mountains and connecting up with the Humboldt River . . . which is the same route as the Hastings Cutoff.

CALEB VOICEOVER: Because this map was published in 1848, two years after the Donner Party trip, I assumed Frémont had simply been trying out Hastings's cutoff for himself. But then, just as I was about to give up and go to bed, something caught my eye.

CALEB IN OFFICE: Wait. What's this?

CALEB VOICEOVER: There, to the east of the Great Salt Lake, I'd spotted a marking on the path. The unmistakable date: 1845.

It seemed that here in front of me was a map showing that someone *else* had taken the very route known as the Hastings Cutoff, before Hastings had even taken it himself. And not just anyone, John C. Frémont, who, Wikipedia informed me, had the nickname *The Great Pathfinder*.

Could it be that Hastings had got the idea for the shortcut after hearing from The Great Pathfinder himself? And if so, what would

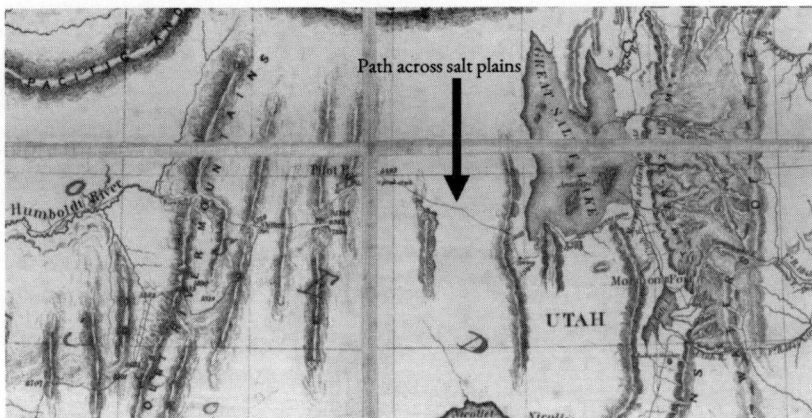

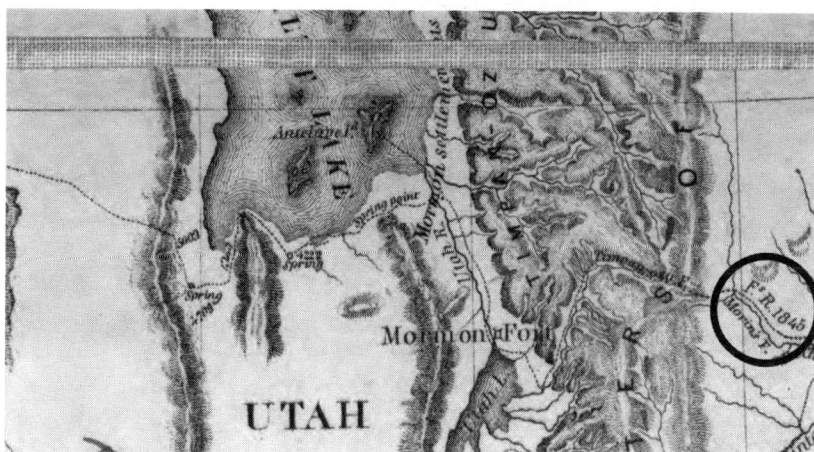

that mean for his legacy as the reckless villain responsible for the tragic fate of the Donner Party?

I was all researched out, and I knew there was someone I needed to speak to.

*　　*　　*　　*

SFX: phone rings.

MRS HOFFMAN (down phone line): I told you to leave me alone, Carter.

CALEB ON PHONE: OK, but Mrs Hoffman, I think I found something.

CALEB VOICEOVER: I told the real Mrs Hoffman about my map discovery, how I'd found source material that could potentially change the narrative on Hastings. After I'd finished she was silent for a moment.

MRS HOFFMAN: You know, Carter, I didn't think you were capable of finding something out for yourself. But fair play. Frémont did take the route before Hastings, you're right.

CALEB ON PHONE: You knew about this?

MRS HOFFMAN: Of course.

CALEB ON PHONE: So, did Hastings ever meet Frémont?

MRS HOFFMAN: We actually know Frémont told Hastings about the route in 1845, when they were both at Sutter's Fort. Frémont boasted that the route took more than 800 miles off the journey.

CALEB ON PHONE: What?!

CALEB VOICEOVER: Not only had Frémont tested the route first, he'd given it his seal of approval, to Hastings. In which case, I had to know—

CALEB ON PHONE: Why is Hastings remembered as the villain? He went off a reliable recommendation.

MRS HOFFMAN: This is how history works, Carter. It's called the Hastings Cutoff. He was the one flogging the thing. You put your name on it, you gotta take the hit. Trust me, no one wants to remember any more detail than that. People want to hear a simple story.

CALEB VOICEOVER: As a podcaster, I didn't need schooling in the importance of a simple narrative – I'd been glossing over inconvenient facts in the pursuit of compelling storytelling my whole professional life.

But I also knew that if anything was more important than simplicity, it was the truth, especially when it came to my own flesh and blood. Something I now felt confident I had found.

And one thing about the truth? It tends to be a little messy.

*　　*　　*　　*

SFX: birds tweeting, wind in trees.

CALEB VOICEOVER: I was walking through Central Park, something I'd literally never done before I started this podcast, because parts of it just felt so *wild*.

But since this story had taken me outside of New York for the first time in my life, even getting close to a real mountain range, the park felt totally manageable, even pleasant. Somewhere I could reflect on everything that had happened.

Yes, I'd recorded a hit podcast and got my first exclusive interview with a dead person – but I'd done so much more than that. I'd discovered I was at the centre of America's most important and famous story, and had successfully rewritten its history for the better.

Lansford Hastings wasn't perfect, and sure, he'd had a part to play in the Donner Party tragedy. But his role was just one of many, including the party's late departure from Illinois, a raging river without a bridge, an early winter storm and a so-called Great Pathfinder championing a lousy path.

What does feel unfair is that Hastings is the one guy who took the rap – as well as all those people who ate each other – and went down in history as a moron.

A label I, as his relative, have also shared.

But the Hastings-Carters are no morons. We're dreamers. And what could be more American than that?

*   *   *   *

As I strolled past a beautiful natural fountain, I couldn't help but wonder what I'd do next. Was there any chance I was related to John Wilkes Booth?

I poured out the rest of my coffee and started spitting into the cup. Because you know what?

It was worth finding out for sure.

*   *   *   *

11

THE MAP THAT'S BLANK (WHERE THE STREETS HAVE NO NAME)

'Hoo-oooh! Hoo-oooh! Hoo-oooh!
Hoo-oooh-ooohhh!'
Bono from U2, 1987

Have a look at this map that we've screenshotted and printed out from Google Maps. It's of a suburb of Kolkata, India, called Bidhannagar.

Do you notice anything weird about it?

(*Map data © 2025 Google*)

It being in black and white doesn't count – colour ink is expensive. In the colour version, the roads are grey, the parky bits are green, the wet bits are blue and the occasional business paying to appear on the map is distracting bright pink. All the usual map fare. But look closer. There's something missing that you'd expect to see in maps of almost every other city around the world. Still not spotted it? Have another look. Here it is again . . .

(Map data © 2025 Google)

Did you spot it this time? The title of this chapter's probably spoilt it, but yes, correct! There are almost no street names.

Before you suggest we zoom in a bit to make the names appear, try it yourself.* You'll see that all zooming in does is make it bigger – this is as detailed as it gets. And if you think Google have been slapdash here, no, they haven't. If you drag the little yellow man to the corner of any one of the anonymous streets to switch to Street

---

* If your fingers flinched to try to enlarge the image before remembering it's a book, don't worry, we've all done it.

View,* you'll see there aren't any street signs on the corner of any of the buildings.

While the lack of street names does make the map look pleasingly sparse and uncluttered, this sea of blank came very close to thwarting the world's most prominent producer of digital online maps.

THE 'NORMAL' SYSTEM OF STREET NAMES

Unless you're on a boat, a plane or a picnic, chances are you're reading this on a street with a name right now. Some examples of street names include Beverley Gardens, Vernon Drive, Baker Street, East Scotland Street Lane, Quex Road and Bwaark Avenue.†

This is a more recent phenomenon than you might expect. Giving streets labels – and insisting we all use the same ones and spell them the same way – only became necessary when the concept of having post delivered to your door came along, which in England was in the 18th century. Before then, people collected their post from designated collection points, such as inns or coffee houses.

But once all streets had names, the system became invaluable, and they became a taken-for-granted part of everyday life. In a lot of cases, the name can provide useful information about the street. For example, the self-explanatory 'Station Road' tells you that the road contains (or contained) a railway station. Sometimes it tells you where you'll end up if you follow it all the way to the end. 'Uxbridge Road' for example is a name used by high streets all over West London, because they all form part of the old road that leads to Uxbridge. Nowadays if you're driving you'd take the A40, but the name's still useful if you want to orient yourself,

---

* Less sympathy if you attempted this, by the way.

† In the first draft of this chapter, this section went on for nine pages. After a lot of back and forth with our editor, we agreed these examples were sufficient.

or imagine how bad the traffic must have been before they built the A40.

Sometimes a street name can give you a quick history lesson on the activity that used to take place there. The unusually named 'Poultry' in the City of London, which today is a skyscraper-lined thoroughfare in the financial district, used to be the best place to buy chickens in medieval times. See also nearby 'Bread Street' (bread), 'Milk Street' (milk) and 'Grape Lane' (*not* grapes – don't google it).

A common trend in seemingly all but the English-speaking world is to name streets after important dates. The world's most famous is probably the main boulevard through Buenos Aires, which has the cumbersome but undeniably memorable name 'Ninth of July Avenue',* making sure no one who visits can forget Argentina's Independence Day.

Another popular convention involves naming roads after important people associated with the area. If you're dead, having a street named after you is a great way to ensure that a part of you lives on for future generations. The very famous Oxford Street in central London, for instance, is named after the no-longer-at-all-famous 2nd Earl of Oxford, who owned the estate it cuts through.†

In more modern times, street names are carefully chosen by developers to tick a number of important boxes, such as ease of pronounceability and spelling for postal workers and emergency services, making sure they don't clash with other nearby streets and that they sound pleasant enough to encourage people to move in. You're more likely to sell a house in Cherry Blossom Lane than Diesel Stain Projects. In many neighbourhoods the street names all fit a theme. If the crescent you live in shares its name with an 18th-century poet, there's a pretty good chance that other 18th-century poets will be lurking around the corner.

---

* It's slightly more cumbersome – by three syllables – in Spanish.
† The fact that if you follow it out of London, going straight on for as long as you can, it also happens to lead to Oxford (via the Uxbridge Road) is entirely unrelated, and merely a gobsmacking coincidence.

'Ah, but what of North America?' we hear you bellow. It's common for cities in that part of the world – many of which had the luxury (or misfortune, depending what flavour you like your urban layouts) of being planned and built in one go in a utilitarian grid form – to simply *number* their streets. The best known example must surely be Manhattan, whose iconic grid (horizontal streets counting northwards, vertical avenues counting westwards, Broadway screwing it all up) is known the world over, enabling tourists to glance up at a sign on any street corner and find their way simply by counting, without ever having to consult a map or interrupt any New Yorkers who are walkin' here. It's very useful for the purposes of navigation or remembering what order numbers are supposed to go in. There are similar systems in hundreds more American cities, from Los Angeles, California, to Monowi, Nebraska. Do these not count as streets without names?

Well, no. While these numeric labels are dull and unimaginative, they are still – importantly for our discussion – *names*. These numbers (and just sometimes for a treat, letters) do still appear on maps and on street signs, and are commonly referred to by people who live and give directions on them, making it ultimately the same system. The point is, the *long thin gap between the buildings* gets a label, and the buildings on either side of it use that label to identify where they are.

But despite this system's usefulness and ubiquity, it's not universal. There are many places around the world, not just India, where tasks like navigation, sending letters and rescuing cats stuck up tall trees are all achieved in streets that have no names.

THE 'NOT NORMAL' SYSTEMS
OF STREET NAMES

In Japan it's the blocks rather than the individual streets that get labelled. This results in maps that look weird to Westerners, where the gaps between buildings where the cars go are an empty void, but the clumps of buildings between them are festooned with numbers. It's a complicated but very clever and well-organised system, resulting in postal addresses that look more like phone numbers.* But like so many other brilliant Japanese ideas, like smart toilets, pod hotels, vending machines for power tools and apple-flavoured KitKats, it hasn't caught on outside Japan. (Except in Korea.)

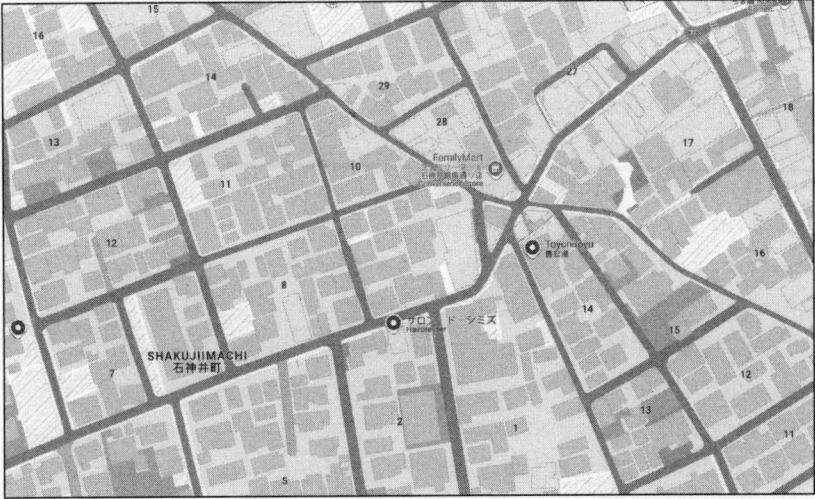

A typical Japanese neighbourhood. What a lot of numbers! Looks more like a dot-to-dot than a map. Don't bother trying, though – you won't draw anything nice, and you'll deface this book/e-reader for nothing.
(*Map data © 2025 Google*)

In most cases, the system of street naming – or lack thereof – is consistent across entire countries. But in Mannheim, a city in south-west Germany, the grid layout in the historical centre (which, for a European city, is already unusual) is labelled not by streets but by blocks, with letters and numbers. It's similar to the

---

* Incidentally, Japanese addresses go the opposite way round to most addresses around the world, with the prefecture first, then the city, then the neighbourhood, then the block number, then the building number, then the floor, then the door. This is definitely the more sensible way round and feels satisfyingly like zooming in from space down to your front door.

Japanese system but, bizarrely, it applies to just one city centre in a country that otherwise uses normal street names. Unless you're familiar with Mannheim's quirk, and you know that the signs on the corners of buildings refer to the blocks rather than the streets, walking around Mannheim can be quite disorientating. Streets can appear to have names that change every few metres, or two names at once. You can turn a corner in Mannheim, and bafflingly appear to have teleported back to the same street you were on a few minutes ago. Hours of fun.

But although these examples are all different from the 'normal' system – the maps of these locations do, at least, provide some sort of system to identify places. In Bidhannagar, however, as well as in much of India, the map is simply blank. But why? Why don't they just fix it? And how on earth does the Indian postal system work?

Let's *address* these questions now.

INDIA

India's system is a bit chaotic. Some streets *do* have names – mainly the big ones. Some streets have 'official' names, but nobody who uses them (the streets) actually uses them (the names). Some have multiple names used by different communities or for different purposes. And some streets have no names at all. This can all be terribly confusing for visitors from the other side of the city, let alone the other side of the world.

This inconsistent system comes about as a result of the fact that Indians tend not to use street names even when they're available. There are lots of reasons for this, but one of them – as is so very often the case all around the planet – is: it's Britain's fault. During the colonial era when the British occupied India, they built a number of major roads and gave them all names for their own convenience and ease of pronunciation, with no consideration for the people who were there first, vastly outnumbered them, spoke a language that wasn't English and had to live/drive/get haircuts on them. It was bad enough that so many of the streets sounded incongruously English (Hill Road, Lamington Road, etc), but it was worse that some names commemorated monarchs who'd never visited and military officers whom most people wished hadn't. The locals understandably declined to use these wildly inappropriate names, going about their business without ever referring to them. As a result, not using street names became something of a cultural habit, which is how you end up with a modern neighbourhood like Bidhannagar. Mostly planned in the 1960s, the developers built rows and rows of streets but didn't feel the need to name any of them.

Another factor encouraging people in India not to use street names is the way its cities have grown. Over the last century, the country has urbanised extraordinarily quickly, resulting in some shoulder-pummellingly crowded neighbourhoods. Many of these are informal or unplanned settlements, better known by the worse name 'slums'. According to their most recent census in 2022, 41 per cent of India's urban population live in slums. In these areas, where webs of narrow, meandering alleyways and dead-ends weave and poke between the buildings, deciding which of these count as streets, or distinguishing where one street stops and another begins, is difficult enough, let alone coming up with names the whole community can agree on for every single one of them.

The net result of all this is that 60 per cent of streets across India have no official name, meaning India contains far and away the most streets without names in the world. And even if they wanted to, it's not like there's any 'quick fix' for naming 60 per cent of India's roads. Because when it comes to India, the word 'big' simply isn't big enough.

It has the biggest population of any country on earth. Its population at the beginning of this sentence was 1,465,130,351 and by the end of this sentence will have risen to 1,465,130,355. Being a country that regularly has to deal with such enormous numbers, is it any wonder that India has its own counting system, with a special syllable-saving word 'lakh' that means 'hundred thousand'.*

A country this size will have more difficulty than most in imposing a nationwide unifying system. Street names don't scale well in a country where there would inevitably be lakhs of duplicate street names to deal with.

Presenting yet more difficulty is the fact that India, as well as being big, is fragmented. The country's 28 states and 8 union territories each have their own distinct culture and history. With 23 official languages and roughly 800 unofficial languages, many of which are mutually unintelligible and with entirely different writing systems, people from one end of the country can have real trouble communicating with people from the other end. It's remarkable that this enormous and enormously varied region of the world technically counts as a single country.

India's federal structure means that each state has a considerable degree of control over how it goes about things, including systematising addresses. So, creating a national standard for street names that could apply across the whole country would be almost impossible, especially as the next president is unlikely to win power on a 'name the streets' campaign ticket in a country with enormous disparities in wealth and high rates of rural poverty.

Perhaps the people all of this affects the most are those responsible for delivering Indian post. Apart from a six-digit 'PIN code' that directs the letter to the correct post office, Indian addresses are an unstandardised free-for-all that could be written in any language, any writing system, and often takes up the entire envelope.

Making matters worse, houses in India – especially in more rural areas – are numbered in the order in which they were built, and not in the order they're positioned in. Looking at the map of a typical Indian street below, number 1 is over there, number 2 is over there,

---

* Get a stopwatch and time yourself saying 'one million'. Now reset the stopwatch and time yourself saying 'ten lakh'. Look at all that time you saved!

number 3 is over there and number 4 is over there. And where's number 5? Over there? No. It's over there. But what about number 6? Well, surely it should be over there. But nope! It's over there.

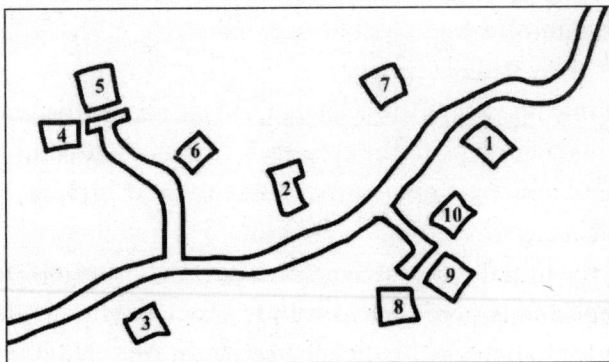

As a basic requirement, postal workers working for India Post have to have a spectacular knowledge of their delivery patch. Think of it as an Indian equivalent of 'The Knowledge', the intensive course London taxi drivers famously have to do to learn all the capital's street names off by heart – only without the street names.

So what does all this mean for mapmakers? Well, in general, mapmakers haven't complained about India's lack of street names. If anything, it just means less work for them.

But then, in the early 2000s, a new type of map came along. One that ran on a set of algorithms, which, in contrast to posties learning the idiosyncrasies of a local beat, relied on standardised data being fed through a set of linear computational processes.

The lack of street names in India was about to become something of a problem.

PRINTING DIRECTIONS

In 2005, Google, which up until then had been a good old-fashioned search engine, launched 'Google Maps', a service that combined digital maps and local data to make suggestions about nearby businesses like hotels and restaurants, and give directions with predicted journey times, both for driving and walking.

Google Maps arrived on the scene during the brief period of history when the Internet was widespread but smartphones were not. So the step-by-step instructions provided to help people navigate from A to B were intended to be printed out, folded up and taken out of your pocket every few metres as you walked (or, if you were driving, blu-tacked to your steering wheel). It was the upgrade to the ubiquitous road atlases that everyone had been waiting for, and people *loved* it.

Seeing the initial success of their Maps product at home, Google saw the enormous potential in rolling the service out across the world, and set about making their website work outside America. For most countries, the website worked pretty much exactly the same way as it did in the US. All that was needed was a quick translation of a handful of useful words like 'turn', 'left' and 'the', and (everywhere except the UK) converting miles to kilometres. But in 2008, when Google Maps launched in India, a typical set of walking directions from Google Maps looked like this:

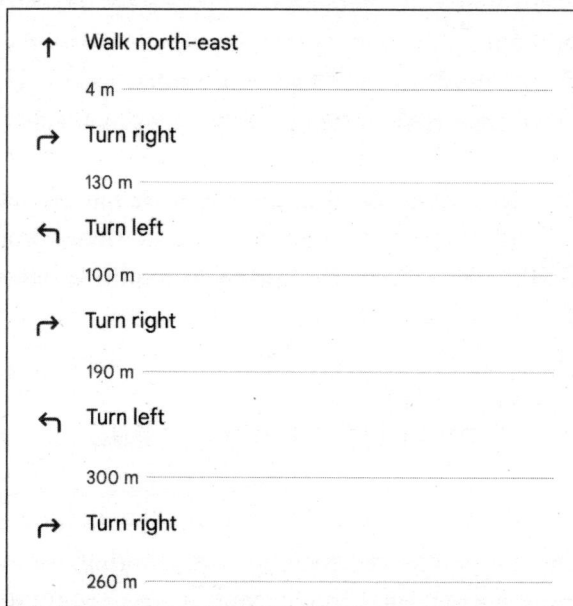

↑	Walk north-east
	4 m
↱	Turn right
	130 m
↰	Turn left
	100 m
↱	Turn right
	190 m
↰	Turn left
	300 m
↱	Turn right
	260 m

Unless you were carrying one of those laser pointers that builders use to measure distances while tutting, or you fastidiously count your footsteps as you walk, you know exactly how long your stride length is and you're really good at maths, these instructions were completely useless. The problem was that Google Maps had been developed by and for Americans. In America, and indeed most of the world, street names are so normal and ingrained, they're taken for granted.

Insiders in the tech industry already joked that the quality of Google's products was inversely proportional to the user's distance from Silicon Valley. But with their product being embarrassingly unusable for more than a billion people in India, they were in serious danger of letting that reputation become well known by regular people outside the tech industry, and, like an Oxo cube with the giggles, they'd become a laughing stock.

Google had several options here:

1. Leave Google Maps exactly as it was and wait patiently for all the Indian states to name every one of their streets.
2. Sheepishly withdraw Google Maps from India, quietly pretending the disastrous launch never happened.
3. Boldly withdraw Google Maps from India, and hold an aggressive press conference demanding that India get its act together.
4. Issue millions of free laser pointers to their Indian customers.

But thankfully they went for option 5. And it was very simple and very clever.

BEYOND THE VALLEY

Olga Khroustaleva was a researcher working on Google Maps at Google's campus in Mountain View, California. Olga, despite spending her days making Google Maps easier to use, was, by her own admission, terrible at map-reading and used to get lost all the time. She had always felt that Google Maps should be easier to use for people like her who had no sense of direction. It should be more reassuring, and it should give directions the way humans do. Her bosses, meanwhile, didn't see any problem. After all, Google Maps boasted the most accurate and reliable routing algorithm in the business. Why fix something that's not broken?

But when one morning in 2008 the news came in that Google Maps was useless in India, Olga spotted an opportunity to convince her bosses that the website deserved some serious tinkering. After all, if they could make the product work better in India, they could make it work better across the entire planet.*

Olga walked across campus (the wrong way at first) and found her way to her boss's door. It opened sideways with a futuristic 'whoosh' noise.

'Olga! What can I do for you?'

'We need to research how people navigate in India.'

'You got it! What room do you need? The one with motion-capture sensors for studying physical interactions?'

'No.'

'The one with the VR headsets for immersive technology experiments?'

'No.'

'The one with soundproof booths for audio and speech-recognition testing?'

'No.'

'The bowling alley?'

'We have a bowling alley?'

'Do you need it?'

'No.'

'The one with the creepy one-way mirrors for observing participants?'

'Actually . . . I think we need to go to India.'

---

* Not to mention cornering one of the world's largest, most lucrative markets.

Unlike many of her colleagues who had just tumbled out of US college, Olga had spent time living abroad and knew that Google had a great deal to learn from the rest of the world when it came to wayfinding. She knew that the only way to properly understand how it was done in India was to hop on a plane and see it for herself.

A few weeks later, Olga and her design-specialist colleague Janet Cheung found themselves on a flight to Bengaluru, India, with two pens and a suitcase full of blank notebooks.

The aim of the trip was to find out how, if Indian maps are all blank, Indians went about the daily tasks that Google's American team had assumed could only be achieved with street names, like putting a venue on a wedding invitation, telling your wife where you parked the car, reporting a pothole or telling the emergency services where you last saw your foot. Clearly, Indian society still functions – life goes on, letters arrive and people know where to meet, but *how?* If they could find out how people in India give directions without using street names, maybe they could teach Google Maps to do it the same way.

They set about their research, using a method that was, in their own words, 'scrappy'.

Instead of collecting vast amounts of data for the computers to crunch, they would find a medium-sized group of real people and talk to them. They gathered (and paid) some participants, interviewed them in their homes and asked them to draw diagrams for familiar routes, as well as keep a diary for a week, writing down every time they gave directions to anyone, including their kids. They phoned businesses, asked for directions to their store and wrote down the instructions they received. And they drove volunteers to unfamiliar places and followed them around, watching them trying to find their way back home.

As their notebooks started to fill up, they discovered what the magic secret ingredient was: when their volunteers described the location of their home, school or place of work, their descriptions all had one thing in common: landmarks.

These landmarks could be almost anything, from a restaurant, statue, shop or petrol station to a man standing creepily on a street corner who mysteriously hadn't budged in years. As long as it was something permanent, easy to see and easy to identify when you walked or drove past, it could be used as a landmark. And these were the elements that made all Indian navigation work.

For example, whenever their volunteers got in a taxi, instead of giving the address to the driver, they told them the nearest landmark and then gave the last few turning directions when they were almost at their destination. The street that the final destination was on – whether it had an official name or not – simply never got mentioned.

To Western ears – especially taxi drivers' ones – this may sound cumbersome and inconvenient, but if you think about it, this is how you actually give directions *most* of the time.

Imagine you're in a Tesco you know really well, and an old lady – mistaking your dark blue shirt for a Tesco uniform – stops you to ask where the crumpets are. While you *could* say 'aisle 5, shelf 13', you're probably more likely to say something like 'Go that way, where the cheese is, on the left, they're in the aisle with the bread, next to the wraps.' You're intimately familiar with that Tesco, but you've never paid attention to the aisle numbers. Why would you? You don't need them and they're never on bloody display anyway.

If you think this logic only applies to crumpets, then pay for your crumpets, step outside into the car park and wait for someone to ask you for directions out there in the real non-Tesco world. As a challenge, try giving them directions without naming any streets. You may find it's surprisingly easy, and that you probably weren't going to say the street names anyway.

And if you found that *too* easy, try giving the next people directions without using the letter E.

Actually, even that's easy. Look . . .

Go straight down this road towards McDonald's. By McDonald's is a road – turn right onto it. Follow that road around, and turn right again. Turn right, turn right, and you'll spot a big roundabout. At that roundabout, you want a third turning. Follow this road straight until you find an additional roundabout. You'll want to turn back around now, and go past an aquarium (which is also known as a zoological fish building). Your final goal is on your right.

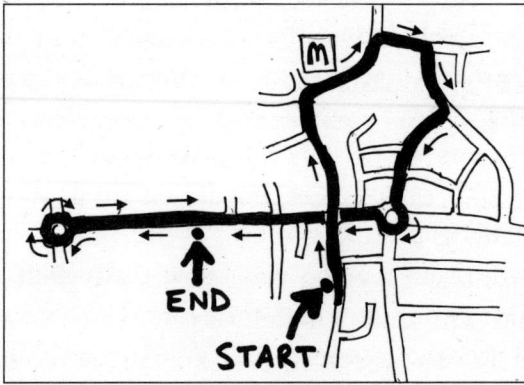

Once Olga and Janet established that Indians make heavy use of landmarks, the next thing they wanted to find out was *how* they use them.

Looking at their notes, they discovered that landmarks didn't just come up at junctions, but at regular intervals throughout the journey and in four different ways:

1. Orientation (head towards this landmark)
2. Description (turn past this landmark)
3. Confirmation (you'll see this landmark)
4. Error correction (if you see this landmark you've gone too far)

So, landmarks didn't simply tell you where to turn; they did another job that was just as important, one that Google Maps had completely overlooked – they *reassured* you that you were still on the right path.

But if Google Maps were going to emulate this, they had another couple of important things to find out: what type of landmarks did people in India use, and how could they build a landmarks database in such an enormous country?

SCALING UP

While it was relatively easy to poll their volunteers in one Indian city about what the most used landmarks were, they couldn't scale this method to every single street in the entire country. Google Maps needed a way to find their own landmarks and know which ones to use. This is harder than it sounds, because something that might seem like an obvious landmark can turn out to be useless to people who actually live near it.

Take the Empire State Building in New York. Huge, so presumably perfect as a landmark, but actually a terrible one, because at street level it's very easy to walk or drive straight past it without noticing.* Even if you're on the right block, you might need to stare at the sky to check where it is – which, in a car, would require opening the sunroof. Not ideal.

Olga and Janet noticed that one type of landmark that came up rather a lot was shops. This meant there was a rather neat solution. All Google Maps had to do was use the businesses already submitted by users and listed in their database as landmarks. This would have the unintended but excellent side-effect that some businesses would benefit from getting extra exposure to their potential customers. And the more users submitted more data, the more the directions would keep improving.

And so, after a further two weeks of talking to locals and driving round and round Bengaluru (the majority of which was

---

* Of course, if Google were making directions for helicopter pilots, this would be a different story.

apparently spent in one particularly nasty traffic jam), Olga and Janet flew back to California with both pens intact and their suitcase full of notebooks now saturated with notes.

When they got back to Google Campus, Olga shoved the notebooks into the funnel of the Slurpshred Codeclankinator™, which made a 'crunch-whoop-shkreeee' noise, and 20 minutes later Google Maps' directions now looked like this:

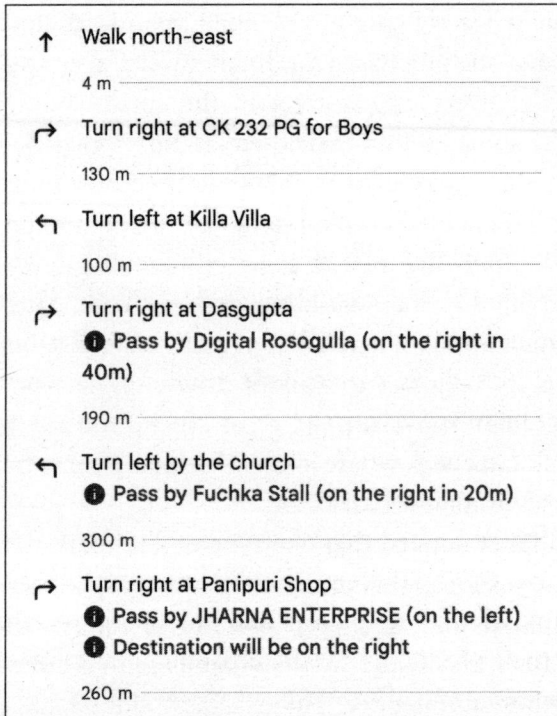

↑   **Walk north-east**

    4 m ————————————————————

↱   **Turn right at CK 232 PG for Boys**

    130 m ————————————————————

↰   **Turn left at Killa Villa**

    100 m ————————————————————

↱   **Turn right at Dasgupta**
    ⓘ **Pass by Digital Rosogulla (on the right in 40m)**

    190 m ————————————————————

↰   **Turn left by the church**
    ⓘ **Pass by Fuchka Stall (on the right in 20m)**

    300 m ————————————————————

↱   **Turn right at Panipuri Shop**
    ⓘ **Pass by JHARNA ENTERPRISE (on the left)**
    ⓘ **Destination will be on the right**

    260 m ————————————————————

The interface was essentially the same, but now it used landmarks just like Indians do, appearing beneath the turning directions in little blue text. It gave the pleasing impression of Google's directions being interrupted in a different font by an Indian guide, piping up with local knowledge in a different voice.

Olga's hands-on approach to qualitative ethnographic research, and Janet's skill with design, fixed the issue, and Google Maps went on to flourish in India. Today, Google Maps is the dominant wayfinding platform in the country with over 50 million searches per day, powering more than 2.5 billion kilometres of directions. And those numbers keep growing at a phenomenal pace.

Arguably, Olga and Janet's research trip had made Google Maps even more intuitive to use in India than it was in America. Olga reminded her bosses of what the purpose of her trip had been all along, to improve the maps for everybody, not just people in countries without street names. She suggested that the landmarks feature should be rolled out globally. And it might just have happened too, but then technology moved forward once again.

WORTH THE TRIP?

By the 2010s the majority of people around the world with internet access got it via their smartphone. Directions, rather than being printed at home, were now followed on the fly. Instead of reading a set of instructions, navigation had become a case of staring down at a device that knew exactly where it was, one that showed the user's location as a blue dot on the screen while the map moved and rotated around it, negating the need to read any words at all.

The problem of how to make Google Maps' written instructions more intuitive was thus simply no longer urgent. Google's engineering team, who, prior to Olga and Janet's trip to India, already considered their product not to be broken, stuck to their guns and insisted even more steadfastly than before that it didn't need fixing. Google's blue 'Pass-by' text today generally only appears in a select few territories where street names are lacking and the service otherwise wouldn't work, such as India, Pakistan, Saudi Arabia and Nigeria. But in Mountain View, California, you could drive right past Google's campus and Google Maps just won't tell you.

It could be argued that Olga's research trip was unnecessary and all Google had to do was sit around for a few years for the problem to sort itself out with satellites. But this ignores a couple of important factors. First, while nobody prints directions on a piece of paper before leaving the house anymore, they do still plan their journey in advance. Understanding where you're going to be is almost as important an aspect of wayfinding as understanding where you are right now. And second, it's unlikely Google Maps could have grown to become the behemoth of usefulness it is today without remaining consistently useful as it steadily grew over the last two decades.

A BETTER FUTURE?

Google didn't sit around and wait for India to develop an addressing system. They pulled their fingers out, found their own hack and got the job done themselves, enabling Indian businesses and their customers to thrive and benefit respectively in the 21st century. So, with Google and its Indian customers having expertly found a way around what used to be a problem, does it matter anymore that India doesn't name its streets?

The sad truth is that there's a serious side to the blank maps of India and serious disadvantages for those living on a street with no name. A lack of a formal address is strongly and consistently linked to poverty. Having an official address is a way of saying that your government knows you exist. Without one, there are all sorts of things Indian citizens can't easily do, like open a bank account, receive a state pension or get the ever-essential Aadhar card, the government-issued ID that lets its citizens access such vital things as pregnancy support, food subsidies and schooling for children.

But whereas as little as 15 years ago the question posed to solve this problem would have been, 'How do we roll street names out to all Indian citizens?', Google's ingenuity has shown that we can cut out that bureaucratic middle and instead ask the more direct question of how, with technology, can we improve the lives of people in India?

This story demonstrates not only how important it is to listen to users when developing technology, but also why we shouldn't take the simplest of things for granted. If something so fundamental as 'streets need names' turns out not to be true, what else could we as a society be doing in a completely different way? Do doors need handles? Do libraries need books? Do babies need nappies? Do cars need brakes?

This story also reminds us to keep an eye on what changes are coming in the future.

With the exponential increase in everydayness of GPS technology, it's not that hard to imagine Western-style street names falling out of use in favour of a system designed for location-sharing with mobile devices. Who's to say that a universal world-wide standard won't come along soon that renders old-fashioned street names functionally obsolete, ushering in a world where knowing the human-pronounceable name of your street is as archaic and pointless as memorising phone numbers has recently become? The transition has already begun sneaking up on us. After all, when was the last time you punched in anything other than the very minimum – a postcode – into your sat nav?

It's not ridiculously implausible that a housing development could be built in the UK in the year 2070 where, for the first time in centuries, the developers decide to entirely skip the step where they come up with names for the streets.

If you're reading this after 2070, and this hasn't happened, write us a letter to tell us we were wrong, and send it to pssssssssschxh-chffffw%%^{*}*%#x▓░░░░░▓░x▓░░░░░ %.

12

THE MAP THAT SHOULD HAVE KNOWN BETTER

*'I find television very educating. Every time somebody turns
on the set, I go into the other room and read a book.'*
Groucho Marx

Tom Pratt has perfected a certain face. It's a face that at once says, 'Relax, there's no judgement here,' while somehow simultaneously saying, 'You are, quite clearly, a complete moron.'

Tom is a British online content creator who has carved out a successful niche tapping into a well-loved stereotype of Americans held by non-Americans: their ineptitude at geography. He exposes this apparent national weakness with cool, calm, straight-faced simplicity. Using video chat sites like Camsurf and Omegle, which connect users with random people from across the world, Tom strikes up one-to-one conversations with various (usually young) Americans, and springs a geography pop quiz on them.

A neatly edited highlights reel of these pop quizzes then makes its way onto TikTok, Instagram, YouTube, or all three, replete with finely tuned facial reactions – kind enough to both keep his companion on the chat, while clearly communicating to his audience what he really thinks of their (frequently confounding) answers.

For instance, in one video he shows his transatlantic companion a picture on his phone of a map of the North and South American continents, with the United States of America clearly coloured in red.

'What country is this?' he asks, deadpan.

'Asia?' comes a first response, which, it transpires, is a not uncommon answer to this question.

'Is it the UK?' asks another, hopefully, as Tom suppresses the flicker of surprise that has crossed his face.

When Tom tells his fellow chatters it was their own great nation of America, responses range from embarrassment to genuine shock.

'But what's that other red bit?' quizzes one.

'That's Alaska,' replies Tom. 'It's a state.'

Another 'test' includes seeing if anyone recognises Australia. This is a level up from Americans recognising their own country, of course, as Australia is decidedly less relevant to Americans than America. But then, it's also a particularly recognisable shape, if you've ever been interested enough to look. In one video his bashful subject is struggling.

'It begins with "A",' Tom helps out.

'Amalasia?' she offers.

He asks another person what they think the smallest country in the world is.

'Florida,' comes their best guess.

It's easy to indulge the American geography-fail stereotype, as we're undoubtedly doing right now; the temptation not to include more unbelievable examples of interactions on Tom's channel is an exercise in self-restraint to which we're particularly unsuited. After all, what better way to bask in the smug self-congratulation of our own basic geographical literacy than watch the feeble attempts of others getting it horribly wrong?

Americans, on the other hand, are understandably tired of the trope. Especially the ones who know their Georgia from their Georgia.* And yet it repeatedly rears its head in popular culture, and as much at home as abroad. For instance, a vox-pop segment

---

* People from the country we call 'Georgia' don't actually call it 'Georgia', they call it 'Sakartvelo'. If you can remember this fact an entire year after reading the book, you have surpassed all reasonable expectations.

from *Jimmy Kimmel Live!* went viral in 2018, in which his team asked Americans on Hollywood Boardwalk to name any country on a blank world map. Apparently, lots were unable to name a single one. Nine years earlier, Jay Leno had also taken to the streets to ask basic geography questions such as, 'What are people from Denmark called?', to which he famously received the answer, 'I dunno, Denmardians?'

Beyond these moments of national stupidity spread across the internet, there have been endless articles, breakfast show segments, scholarly papers and everything in between all dedicated to asking the same question, 'Why are Americans so bad at geography?'

But first, are they? Because – believe it or not – a talk show host asking a few random strangers to exchange their dignity for a momentary appearance on television doesn't prove that all Americans *are* bad at geography; neither does it prove that they compare badly with other nations. All it proves is that *some* Americans are clueless. Which, in a country of 340 million people, is no great news story.

Stereotypes are hard to shift. Once we believe something to be the case, and it feeds into a wider idea that somehow makes sense to us, such as 'Americans are too inward-looking to be good at geography,' we become hard-wired to only see more and more examples proving the phenomenon's existence. It's a trick of the mind commonly known as 'confirmation bias'.

No doubt you could easily find plenty of Brits, Germans, Kartvelians* or Denmardians who'd be similarly clueless and package their gobsmacking answers into an equally compelling montage, if you only asked enough people the same seemingly simple question. Because, of course, what the likes of Tom Pratt, Jay Leno and Jimmy Kimmel will never show you are the many, many Americans who answered the questions posed to them with carefree ease. That is, quite simply, not good television. Or TikTok.

* People from Georgia (the country).

Or book. Which is why it's a relief (for us) that the stereotype is supported by some statistical evidence. The most notable (although increasingly not all that recent) being the *National Geographic*-Roper 2002 Global Geographic Literacy Survey that placed America second last out of nine countries surveyed.* One standout statistic from the survey was that of the 3,000 18- to 24-year-olds questioned, one in ten couldn't locate the USA on a world map. The report opened the floodgates for an outpouring of concern – both nationally and internationally – about American geographical skills.

Another more recent survey, in 2017, tested 2,000 respondents – half from Europe and half from the USA – on their ability to accurately locate 16 countries from across Asia, Africa and South America. The percentage of Europeans who got the answer right was higher than Americans on all 16 of those countries, and on average they were 8.7 per cent more likely to name them correctly. One way of reading this is to conclude that the results weren't *wildly* different from one another; at the same time, if 8.7 per cent were a guaranteed return on a financial investment, you'd rightly think, 'That's massive!'

But what *really* doesn't help the perception of Americans as badly geographically challenged is that their geo-blunders are not limited to individual survey subjects; they've very visibly spread to one of the nation's supposed bastions of trust, the news. Because while one can certainly understand a number of uninterested individuals confusing their Africa with their Asia (and we'll look at some reasons for this sad geographic apathy† shortly), what is much harder to understand is when major national news networks show the same astounding level of dunceness. And it's something that happens with surprising regularity.

---

* GB came third last. Mexico came bottom, Sweden top.
† Geograpathy.

FAKE NEWS MAPS

Had you tuned in to CNN on 5 October 2013 you'd have been 'treated' to a displeasant story about a deadly hornet outbreak in Hong Kong, China. The producers had wisely decided to include a big, flashy, animated 'locator' map, the name given to a simple graphic that provides geographical context to the story.

As the presenter introduced the news item, a large blue globe spun on the giant screen beside them, finally coming to rest in . . . South America.

Huh?

Maybe the animation glitched. But no. There, on the Brazilian coast, pretty much exactly where you'd find Rio de Janeiro, the label 'Hong Kong' has been clearly written in large, strident letters. The only label on the entire map.

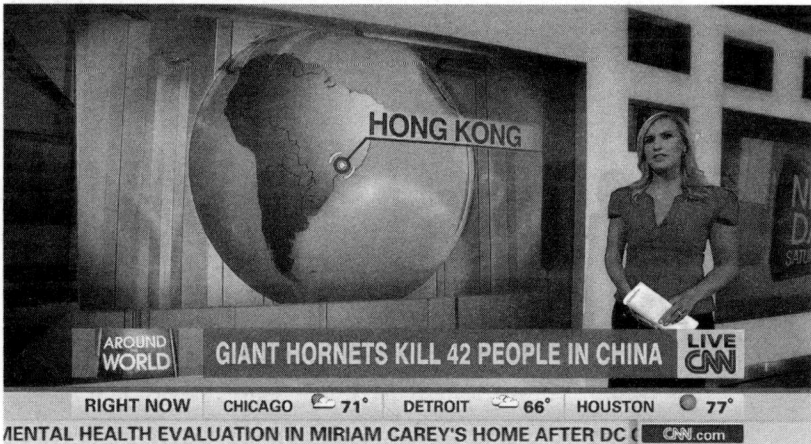

The map was widely shared and ridiculed on social media, and quite rightly so. After all, CNN is the sort of network that prides itself on the strength of its international reporting, certainly compared with some of its national news network competitors.

In many ways, it's the perfect map mistake. It's a map with only one purpose – to label a single city on a map of the world – and in that it has exquisitely failed. And it's not like Hong Kong is an obscure city; home to 7.5 million people, it's one we regularly hear about in the news for being politically at odds with the Chinese regime. In fact, a decent number of primary school children would probably be able to tell you what country it was in, so one would certainly expect the Cable News Network to correctly label its position on the coast of southern China, or at least place it *some-where* in China. Or Asia. Or the Northern Hemisphere . . .

No. Instead the one adornment to this strange globe-map (where South America is so big that Venezuela is encroaching the Arctic Circle) is the enormous, capitalised words 'HONG KONG' in possibly the funniest place they could have written them. Its comedy lies in its confidence. 'IT'S HERE!' screams the map. 'IT'S #£%@ING NOT!' you scream back.

If we were to attempt to defend the map at all, we could say it was probably only included in the news item to do the specific job of reassuring the audience that 'these hornets are nowhere near *you*'. So, unless anyone was tuning in from Brazil, it very sort of did its job.

Sadly for CNN, the Hong Kong fiasco was far from their only slip-up. It appears from a variety of online screengrabs and photographs of people's televisions that CNN also managed to produce maps that:

- misplaced Israeli city Tel Aviv in the Golan Heights to the north-east of the country, rather than on its western coast.
- confused France with Italy in an item about coronavirus.
- put the Libyan capital Tripoli in Syria for an item about Colonel Gaddafi.
- suggested that Ukraine is in Pakistan.
- located London in Norfolk, 120 miles away from London.

CNN might appear to be the most regular offenders, but they're certainly not the only American network to have suffered embarrassing map misfires of this sort. In a 2017 report about a terrorist attack in London, MSNBC located Vauxhall, an area of central London, 35 miles to the north of the capital, in Luton. Similarly, in 2009 a Fox News report about the Middle East showed a map that mislabelled Iraq as 'Egypt'.

Perhaps surprisingly for a network of a similar size to CNN, this is practically the only Fox News map blunder that's been widely shared online*. Either Fox News have got the Geography Bee winner on their team, or perhaps they simply produce fewer news segments about the world beyond the USA. Who knows!?

'Displeasant', by the way, in case you're still thinking about it from several paragraphs ago, is not a word, but it should be.

Americans will be keen to point out that American networks are not the only culprits capable of cartographic crimes. The BBC once put Romania in Bulgaria, Sky News put Sunderland in Cumbria, Cuatro al Dia (Spanish) put Madrid in Salamanca. Another infamous howler came from Russian network RT (formerly Russia Today) in a piece about potential missile bases in Japan, South Korea and Australia. RT decided to centre the story's map on Australia, but, seemingly keen to show the two other countries in the story without the faff of zooming out, somehow decided that Papua New Guinea could stand in as 'South Korea', while New Zealand would make do as 'Japan'.

But before our American friends start feeling too relieved, a bit of digging reveals that although this particular mistake occurred on a Russian-owned network, RT had established a number of spinoff channels overseas. And this particular error was produced by none other than RT America, headquartered in Washington, DC, and mostly staffed by American citizens.† We cannot say for sure who was responsible for this particular blunder, but chances are it wasn't anyone from Russia.

---

* Bar one other, where they somehow mistook Japanese nightclub 'Shibuya Eggman' for a nuclear facility.

† RT America stopped broadcasting in 2022 because Russia became … busy.

Screenshot from RT America, broadcast in 2019.

What RT did do well, however, was apologise, issuing a state-ment in good humour that ran:

> Our American early morning news team suffered a little geog-raphy mishap [. . .] [We] have corrected ours as soon as it was spotted, and have given our team a new map of the Southern Hemisphere to ensure it doesn't happen again.

Good for them, although it's unlikely that the reason for this error was the accuracy of their *old* Southern Hemisphere map. Which begs the question: how are news networks getting graphics like this so horribly wrong, and why is it happening so often?

WHY SO MANY MISTAKES ON TV NEWS?

For the most of us who have never worked in broadcast news, it's hard to imagine just how time-pressured an environment it is. In a competitive network landscape where all news providers are vying for viewers, speed is critical. You want to be either breaking, or as close to breaking, the Big New Juicy Scandal as possible. You can't dither on about yesterday's news while everyone else is revealing new information on the BNJS.* Decisions are made fast, with reports being put together by brilliant minds making dozens of editorial, compliance and legal decisions often just a matter of minutes before the next bulletin goes out. And, as you'll know if you've ever tried making a mojito while running for the bus, when things are rushed, things go wrong.

There are as many different excuses for map mistakes as there are map mistakes, but often it comes down to a combination of human error and tricky software. For instance, an editor may look at a locator map and suggest that more regional context is needed, asking their graphics employee to zoom out a bit further. If the labels have already been applied as a separate layer to the map and they aren't readjusted after zooming, they'll now become misaligned. If that last-minute tweak doesn't get properly checked by someone with the geographical expertise to spot that something's gone wrong, the mistake could easily end up being broadcast.

This might explain how London ended up in Norfolk on CNN's map. (We can't say for sure that's specifically what happened, but they'll probably prefer that interpretation over 'CNN don't know where London is.')

Another explanation is the fact the news networks often recycle old maps. When a map is shown on the news it's not created from scratch every time. Graphics departments have vast libraries of maps and templates that can be reused, or adjusted slightly when needed, providing a prime opportunity for problems.

The graphics bod could, for example, use a template from an old segment about Rio de Janeiro, then get as far as replacing the text, but forget to shift the globe base. This might explain how CNN ended up with 'Hong Kong' in South America.

Sometimes it's a simple case of getting two places with the same name mixed up. This one comes up rather a lot.

* Not an official news acronym.

Take for instance CNN's map that misplaced the Libyan capital Tripoli in Syria. There *is* a Tripoli in Syria, but it definitely wasn't where Gaddafi was hiding at the time, and nor were CNN intending to suggest so.

Similarly for MSNBC's map that put Vauxhall in Luton, the mistake becomes understandable when you know that Vauxhall car manufacturers had an enormous and very famous factory in the exact spot they chose to annotate Vauxhall. It doesn't *excuse* it, but it does at least *explain* it.

Then again, on that same map there were other errors that we cannot fathom an excuse for, such as the label for London Bridge being placed in Enfield in the far north of the capital's suburbs, rather than on the River Thames where London 'Bridge' (rather obviously) is.

The truly inexplicable ones are anybody's guess. Could it be sabotage from a disgruntled employee, or the graphics person simply had a very bad day at the office? Or is an idiot? One tempting interpretation is that the news networks are in fact getting the maps wrong on purpose.* Unlikely though this is, there's evidence to suggest that it's not a wholly bad idea…

OUTRAGE

There's something about seeing a place one knows very well being misrepresented on a map – particularly if it's the place one's from – that can induce an uncontrollable spitting of one's tea from one's mouth across one's drawing room, potentially ruining one's Persian rug (should one own one). And, of course, pursuant to rug cleaning, unlocking one's phone and posting about said map online.

Because one of the sad facts of the social media age that has become so familiar is that outrage = engagement. And engagement = clicks and free promotion. As a result, generating outrage

---

* See our similar speculation in Chapter 1.

has become an entire industry, one that many news outlets around the world (if not most) have become skilled at harnessing.

It's no great secret that news networks manufacture outrage in the way a news story is selected or presented to increase clicks. In the digital age, news businesses on both the left and right have realised that stories provoking strong emotional reactions are more likely to yield a financial return either through advertising revenues or more exposure leading to increased subscriber counts. As a result, any story that presses people's peeve buttons is of value. And, to be clear, we're not suggesting that CNN would be interested in generating clicks by putting Hong Kong in Brazil. As a brand that presumably aims to align itself with trust, loyalty and responsibility, only a deep conspiracy theorist with a trademark blind spot for basic human reasoning is going to suggest that CNN would do something that makes them look so inept in exchange for a few hundred likes.

Nonetheless, creating bad maps on purpose might be the sort of thing some companies would be tempted to do to drive engagement and publicity. A map of the UK produced in a pamphlet by energy supplier Octopus Energy in 2024 labelled 16 cities across the United Kingdom, getting about four of them close enough to their actual location but twelve of them bizarrely off, and all in different directions to where they should be – some east, some north-west, just all over the place. Again, we have no idea how, or why, this happened, but the fact that a tweet complaining about it by a user called @geekcubed was viewed more than a million times suggests that, whatever the motivation, Octopus Energy probably looked back at that as a decent bit of brand exposure.

Another tweet, this time by beer manufacturer Carlsberg from the handle @CarlsbergUK, also in 2024, used a map of the UK to show pints of beer raining down as a form of relief from a 'hot summer of football' on the horizon. Bizarrely and inexplicably, the inlet on the south-east coast where the Thames Estuary drains into the North Sea had been extended all the way into London. As a result, instead of green land where London should be, there was a big, blue, watery blob, as if the Thames Barrier had taken storm season off and the entire city had been submerged (way in advance of even the bleakest of climate change warnings). What at first glance had appeared to be an ordinary Carlsberg advert suddenly had become something to comment on, retweet and be shocked by.

Whatever the intentionality behind a bad map, one thing's for sure: people will talk about it, share it, laugh at it and generally let everyone else in on the fact that they know something the map-maker didn't. Interestingly, of all the traditional school subjects, it's errors in geography that seem to be among the most enjoyable to point and laugh at. Get a history date wrong? Forgivable. Grammar all over the place? Shrug. Struggle with arithmetic? Relatable. But expose your terrible geography by misplacing a location on a map? That's a cognitive crime it seems we're not ready to forgive.

And, if we're going to accuse America of foul play, the least they deserve is a fair trial . . .

GUILTY?

UNITED STATES SUPREME COURT, NEW YORK CITY

CASE NO. 978-0-00-871027-9

THE PEOPLE OF THE UNITED STATES (Plaintiff)

vs.

ALL AMERICAN CITIZENS WHO THINK AFRICA IS A COUNTRY (Defendants)

ADDRESSES: BOTH VARIOUS AND UNKNOWN

TRANSCRIPT OF PROCEEDINGS

CLERK: Calling THE PEOPLE OF THE UNITED STATES vs. ALL AMERICAN CITIZENS WHO THINK AFRICA IS A COUNTRY – some of whom we believe are still trying to find the building.

All rise for Chief Justice Fairman.

CHIEF JUSTICE FAIRMAN: Good morning. Please be seated and we can make a start.

CLERK: Appearances, starting with THE PEOPLE.

MISS GRATICULE: For THE PEOPLE, District Attorney Sierra Graticule.

CHIEF JUSTICE FAIRMAN: Good morning, People. And for the defense?

MR SPAT: Frank Spat, your Honor.

CHIEF JUSTICE FAIRMAN: Thank you. This court is here to hear THE PEOPLE'S case that all persons who think the continent of Africa is a country are bringing America into disrepute, because geography is an important subject that we should value.

Now, who wants to go first? Hands up. OK, Mr Spat, you <u>were</u> first, but I don't think I'd finished answering the question, so, in accordance with court etiquette I will allow Miss Graticule to speak first.

MR SPAT: Urgh.

MISS GRATICULE: Thank you, Your Honor.

Africa is a continent. It is big, diverse and home to
54 countries. It's a shame on America that so many of
us lazily group these different cultures and
histories into one nation state with such regularity.
But perhaps it should come as no surprise, because
it is symptomatic of a wider issue – that many
Americans are woefully poor at geography. It's a
reputation the nation has struggled to shake, and
we should be ashamed of that.

Geography matters. Time and again our misunderstanding
of the geography of a place has led us into trouble,
from our ill-coordinated response to the damage caused
by Hurricane Katrina, to the protracted war in
Afghanistan, where we failed to appreciate the
challenges of both topography and political culture,
to that time Joe Biden thought Glasgow was in England.

CHIEF JUSTICE FAIRMAN: Haha, yes, that was good.

MISS GRATICULE: No, Your Honor, it wasn'—

CHIEF JUSTICE FAIRMAN: Counsel!

MR SPAT: Thank you, Your Honor.

It's convenient for the world to think of Americans
as 'dumb', isn't it? The world's most affluent,
confident, powerful nation, whose freedoms are the
envy of many. For those not fortunate enough to call
themselves American citizens, it makes sense that
they look for something to criticize, to make them
feel better about <u>not</u> being American. That thing,

it appears, is whether or not a handful of people can identify all the countries in the world.

MISS GRATICULE: Objection, Your Honor. The issue at hand is whether Americans can name a <u>single</u> country.

CHIEF JUSTICE FAIRMAN: Sustained.

MR SPAT: The truth that the rest of the world doesn't want to accept is that Americans are smart. Our average IQ is higher than it is in two-thirds of European countries. Our contribution to global technologies, innovations, film and literature rival or surpass any modern nation. Our research institutions are the envy of all. Americans recognize what is important in the world—

MISS GRATICULE: Objection, Americans do not even recognize the world.

CHIEF JUSTICE FAIRMAN: Overruled, there's no evidence for that.

MR SPAT:—and we excel in what matters. We have nothing to apologize for.

CHIEF JUSTICE FAIRMAN: Not bad. State? Any thoughts?

MISS GRATICULE: The counsel for the defense seems to forget, Your Honor, this is not simply about what other countries think. Many Americans are themselves mortified at our national ignorance, laying the blame squarely on an education system that has underserved our children in teaching them about the world we live in.

Ever since Harvard made the catastrophic mistake
of removing geography from its course list in 1948,
many other universities have followed suit. When
geography _is_ taught in our school classrooms –
which is perilously rare – the requirements are
loose, with hours often filled by US history,
civics or economics instead.

MR SPAT: And so they should be! We are a large
nation with our own rich and varied cultures,
climates, landscapes and current affairs between
our many states. Preserving American Greatness is
about taking care of our own house.

Americans learning about America is not only
valid, it's preferable. Our meddling in foreign
affairs, our attempts to cultivate world peace and
spread our values over previous decades, were naïve
and misguided. Other countries don't want or need
our help. Let them worry about them, and us worry
about us.

MISS GRATICULE: Your Honor, this is patently not
true. Many countries look to us to lead, and to help
bring about a more just world. [Let the record show
that at this point Counsel Frank Spat spat.] If we
deign to see ourselves as a superpower, we must act
responsibly as one. After all, we cannot consider
ourselves a power in the world if we're not even
willing to engage with it.

International affairs affect America too, whether
it's through trade, migration or threats to our
national security. Sitting by is simply not an
option, and, if we do, the world will revolt.

CHIEF JUSTICE FAIRMAN: You both make a series of salient points, and now, in accordance with federal law, it is incumbent upon me to make a judgment of guilt out of ten, one being not guilty, ten being <u>really</u> guilty. Just give me a moment to think . . .

OK. I've made up my mind. All rise . . .

*

[Let the record show that as the court rose, CHIEF JUSTICE FAIRMAN fell into a deep sinkhole that suddenly opened beneath his chair. He survived the fall, but his gavel fell down after him, the handle impaling his right eye. Everyone immediately forgot about the case, because this was a far bigger news story.]

*

AT A LOSS

Undoubtedly, the perception that Americans are terrible at geography persists, but this fact alone begs a bigger question . . . Do we actually *need* to know where places are?

As democratic societies in a globalised world, a basic understanding of the population centres, power bases and unstable regions feels, at the very least, critical to voting not just responsibly but in our own interests. Yet many Americans are being let down by an entrenched national belief that geography – most broadly, the study of other places, where they are and what they're like – simply doesn't matter. And when one of your country's biggest and most international-facing news networks is capable of producing such laughable map gaffes, what hope does the average Joseph have?

In articles on the subject one common defence of geographical ignorance comes up time and again, and not just in relation to America: in an age of smartphones, who cares if you don't know where something is? *You can always just look it up.*

It's a bleak and specious argument.

It's true. You could. But for one, you probably won't. CNN clearly didn't. And for two, access to knowledge is, by no stretch of the imagination, the same thing as *having* knowledge. After all, it's not like we look up the letters of the alphabet when we come to read a book. We learn them by rote. And that's incredibly useful when reading and writing, which would be arduous to the point of its own oblivion had we not done so.

When information is learnt it is also synthesised, understood and on hand to use at a moment's notice when we need it. What's more, relying on looking something up also requires the knowledge that you actually *need* to look something up, because you can't exactly search for what you don't know is there. It's not like you're going to look up a country in Africa if you think Africa itself is a country to begin with.

Search engines, blue dots on maps and the idea that we don't need to learn things are all part of a troublesome trend of farming out our faculties to devices, while we as a species slide listlessly towards a sort of inert helplessness. For anyone who hasn't seen the Pixar animated film *Wall-E*, it presents a dystopian vision of humanity's future that feels worryingly prescient. The backdrop for this heartwarming mechanical love story is a world where all our needs are catered for by robots and other technologies, leaving human beings all moving about on floating chairs unable to move, think or act for themselves. Their only concern is for their continued consumption, as they've lost all care for anything that's actually happening in the world around them. Is there a more realistic future scenario for our species across the entire canon of End of the World stories?

Technology has done all sorts of amazing and useful things; that's why we keep spending money on it. But it appears to come with the barely hidden risk of human devolution at the same time. Walking may be harder than driving, navigating trickier than being told where to turn and learning more difficult than not learning. But it's still better than the end of all human capability.*

* Sorry to end on a bit of a downer. Here's a fun fact to cheer us all up: did you know, from Season 4 of *Frasier* onwards, the part of Eddie the dog was played by the original Eddie the dog's son?

13

THE MAP THAT BROKE
THE FRAME

*'The younger brother must help to pay for the
pleasures of the elder.'*
Jane Austen

Unless you're a medieval historian, or otherwise particularly interested in the Wars of the Roses in England, it's unlikely you spend much of your time thinking about the year 1461.

Other 15th-century years deserve much more attention. For instance:

- 1415: The Battle of Agincourt, in which England win an unexpected victory over France in the Hundred Years War
- 1453: The fall of the Byzantine Empire
- 1492: Columbus sails the ocean blue

But 1461? As a sequence of integers, it's more likely to be written down as a playful alternative to one minute past three in the afternoon than be considered a historical year of any particular consequence.

On a personal level, however, it was a year of *some* consequence for nine-year-old Christopher Columbus's mother, Susanna Fontanarossa, because she gave birth to another child, although the child was her middlest as it would turn out – the third of five – thus lessening the significance of the year even for her.

This third child was a boy named Bartholomew Columbus.

Bartholomew, of course, was the English spelling. In Genoa, where he was born, his name was written *Bertomê*, and in Portugal – where he was often based – it was written *Bartolomeu*. It might seem unnecessary to dwell on the various international versions of his name, but in truth it's only fair, because from now on, we will only refer to him by his correct name: 'Columbus's Brother'.

Of Columbus's four siblings, three were – in fact – brothers. The other two Columbus's brothers either died too young or were too meek to leave their mark on the world. And so, for the purposes of history, there was only one Columbus's brother, and that was Columbus's Brother.

Columbus's Brother might not, it turns out, have even been born in 1461 (after all that!). Nobody's exactly sure when it was, and Columbus's Brother had no birth certificate. We know when Columbus was born. Obviously. That was 1451, no doubt about it. But Columbus's Brother? Roughly a decade later. Give or take.

As far as we can tell, no books have been written about Columbus's Brother (let alone Columbus's Brothers), but Columbus's Brother should probably be grateful that so many books have been written about his brother that he has nonetheless been mentioned a fair bit. Because from the start, Columbus's Brother's destiny was always bound up with that of his older, bolder and infinitely more famous globe-trotting brother, Columbus.

And what better way to serve Columbus's ambitions than to become what Columbus's Brother became: a cartographer. And a very good one at that. Yet, at the same time, one who wasn't above the idea of completely distorting the shape of the known world on his maps if it meant helping his brother get the patronage he so desperately needed to pay for a voyage the wrong way round the world.

WONKY WORLDS –
A THOUSAND YEARS OF BAD MAPS

Columbus's Brother was making maps at a particularly interesting moment in map history, because in the late 1400s cartography in Europe was on the cusp of a much-needed Renaissance. For the few hundred years pre-Columbus's Brother there had been three main types of map in Europe.

1. Ecclesiastical

Such maps related to the Christian church. While many of these were spectacularly imaginative, they were – practically speaking – useless. Unconcerned with either proportional scale, relative distance, shape, empirical evidence or simply where places actually were, they were instead a heady mix of odd-shaped continents and pictograms of important settlements broken up by rivers and seas that were often about as big as each other. These 'places' were then chaotically woven together with Bible stories from over a thousand years earlier, creating a sort of semi-geographical religious-history diagram with dragons around it. So, on a world map, Rome, Jerusalem and Babylon might all be marked, but so too would Noah's Ark, the Garden of Eden and that lamb Abraham found in a bush and sacrificed instead of his son – as if it were on the way somewhere.

2. The Portolan Chart

The clue is in the 'port' bit, because these were nautical maps. Portolan charts first turned up in the 1200s as compasses arrived and the sea became increasingly important for both trade and warfare, and they were the best there was for getting you from A to B on sea. There were no lines of longitude or latitude, and no cartographic projections to help account for the curvature of the earth. Instead, they comprised a series of lines criss-crossing all over the map, emanating from 16 points of the compass in various locations. In short, they helped you follow a handful of straight lines from a series of chosen points, which, if you were lucky, might include either the point you were setting off from or even going to.

Many ships sank.

3. Ptolemaic

The clue is in the 'Ptolem' bit, because these were maps based on the work of Claudius Ptolemy from around AD 150. And Ptolemaic maps are important, because they form the bridge between the old (rubbish) and the new (less rubbish) maps of the Renaissance.

They're also, in many ways, a paradox, because despite the prevalence of the term 'Ptolemaic maps', none of Ptolemy's own maps have, in fact, survived. In even more fact, we're not actually sure he ever made any – common consensus is that he probably didn't. What he did do was write an eight-book series called *Geography*, none of which, again, survives.

Thankfully, it *was* seen and copied by others, and the earliest *Geography* copy we know of was written by a Byzantine monk in the late 13th century (over a thousand years later), who may or may not have changed or added some bits. Still, better than nothing. And in fairness, it doesn't appear that anyone in Europe had got any better at map-making between 150 and 1200 anyway. In fact, they'd got worse,* so there probably wasn't much to update.

Ptolemy's *Geography* left an indelible mark on mapmaking for centuries for one simple reason: it included a comprehensive 'how to' guide to mapmaking, which insisted on the importance of maths in drawing maps. Ptolemaic maps were the first to include lines of

---

* Hence the term 'Dark Ages'.

longitude and latitude, as well as a mathematical projection for those lines to fairly map locations from a spherical globe onto a flat surface (a map). Having written this guide to mapmaking, Ptolemy also tabulated the co-ordinates of 8,000 known places based on various people's travels, all of which were accessible to him in the historically unparalleled library in Alexandria, where he lived.

Using this guide and these wide-ranging geographic references, various later scholars (and Byzantine monks) drew what have become known as Ptolemaic maps, many of which have survived (see below).

Ptolemaic maps are, to a modern eye, desperately incomplete due to the smaller size of the known world 2,000 years ago (and for centuries after). Major shortcomings include: the absence of America, Australia, southern Africa and the polar regions; the Indian Ocean being depicted as a lake (with the base of Africa linking speculatively up with eastern Asia); and Sri Lanka almost matching the size of India.

Still, I think we can all agree, not bad for AD 150. And yet, Ptolemy's critical innovations were to spend over a thousand

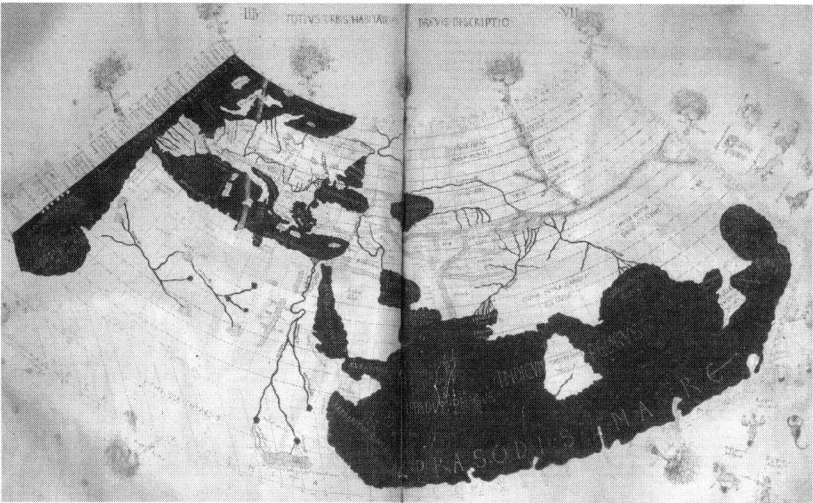

World Map According to Ptolemy, c. 1450–75, credited to Francesco di Antonio del Chierico, courtesy British Library.

years being ignored by European mapmakers until *Geography* was finally translated into Latin in 1406. Only then did he really start to catch on and the European cartographic scene start to light up.

By the time Columbus's Brother was plying his trade in the 1480s, Ptolemy was all the rage. The challenge now was to improve Ptolemaic world maps by journeying further into the world and finding out where more things were.

This was the dawn of the cartographic renaissance in Europe, one that combined the ancient but only recently understood Ptolemaic principles of mathematical cartography with the geographic spoils of voyage and discovery. And no double act was better placed to propel this cartographic revolution than the Columbus brothers.

COLUMBUS'S BROTHER'S MAPS

There are none. Sorry.

While Columbus left behind a sizeable (if not contested) legacy, Columbus's Brother left no material relics of his mapmaking prowess whatsoever. So how do we know anything about his maps and their wrongness?

The answer lies in the clues he left behind in the historical record; in particular, on two particularly famous works that not only drew the world, but changed it. Contained within both are all the hallmarks of Columbus's Brother's knowledge, skill and – above all – agenda.

One of these was a globe. The earliest surviving globe in the world, as it happens, made by a man named Martin Behaim in 1492. The *Erdapfel* (to give its proper name) is more commonly referred to as the Nuremberg Globe. It was made in Nuremberg, where it's now on display should you ever want to see it. And you might, because it's hard to show you a globe on the page of a book, the crux of many a map dilemma.

The brave thing about making a globe rather than a map* is the confident suggestion that you can account for the entire spherical earth. Ptolemy had only ever attempted to describe 180 degrees (half) of the earth's circumference, stretching from the Canary Islands just off West Africa to the Malay Peninsula and Korea in the east. Even then, he'd stretched what is in reality just 115 degrees' worth of the Earth up to 180. But Ptolemy, knowing he didn't know what he didn't know, decided it was best not to guess what might be in the other 180 degrees of the world that nobody from Europe had yet explored.

Behaim, on the other hand, decided he could.

In a bid to fill the missing 180, he began by nudging the Cape Verde Islands in the Atlantic 30 degrees west of Lisbon, when in reality they're only 15 degrees west. He then added Cathay (northern China) onto the east of the Ptolemaic world map, using accounts from the travels of Marco Polo in the late 1200s that hadn't been available in Ptolemy's time, but which Behaim could access in Nuremberg. This added a useful further 57 degrees to the other side, meaning he now had 267 degrees accounted for.

Behaim had now created an enormous Eurasia over 100 degrees wider than it was in reality. But he still needed another 93 degrees to complete his sphere.

Next he added Cipango (Japan), also described by Marco Polo, placing it another 25 to 30 degrees off the coast of China. Nearly there. Behaim had now added 110 degrees on top of Ptolemy's original (massively stretched) 180, and the remaining 70 could simply be made up of the ocean between Japan to the west and the Cape Verde Islands near Europe to the east.

That ocean now looked easily surmountable, especially as there were some perfectly spaced island stepping stones between the two continents made up of the Canaries, Cape Verde, Japan, and the two islands of St Brandon's and Antillia (which didn't actually

* See Chapter 6.

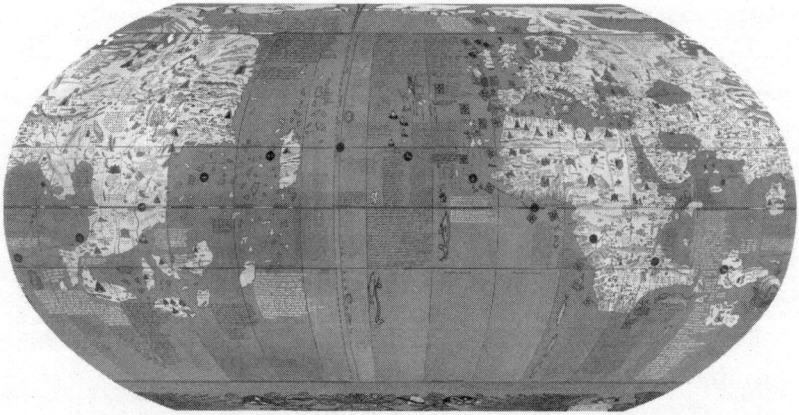

1908 scale recreation of Behaim's Globe by Ernest Ravenstein,
showing the distance between Europe and a misshapen East Asia
(we'll come to that).

exist, but they didn't know that yet)*. The upshot was that the
world Behaim had drawn made the Atlantic route to the highly
coveted riches of China, India and the Spice Islands look like pure
common sense.

But where did Behaim get his ideas? Did anyone have anything
to gain from such a world geography?

A BAD PLAN

It's easy to confuse the fact that Christopher Columbus achieved
something historic with the notion that he was right. Because for
all the praise Columbus has received for his bravery, vision and
seamanship, there's no way around the fact that he was one of the
wrongest people ever.

Columbus had a totally wrong view of how the world looked,
and he was really stubborn about it too. He believed that every

---

* The name 'Antilles' was subsequently used to describe the real islands
of the Caribbean, which weren't far off where Antillia had previously been
imagined.

degree east or west was far fewer miles than it actually was, and he became nothing short of obsessed with the idea that 'the Indies' would be better reached by sailing west. He based this wrong belief on:

- a wrong map by an Italian astronomer called Toscanelli, who massively underestimated the circumference of the earth.
- chatting to some locals in the Canary Islands who said, 'We think there's something that way,' which Columbus assumed was Asia (and was wrong).

Despite all the dead arms Columbus had administered over the years, Columbus's Brother looked up to Columbus like an older brother and unquestioningly agreed with his world view.

In the early 1480s the two Columbus brothers hatched their unique and terrible plan to reach the riches of Asia by going in the opposite direction to where received wisdom said they should, naming their project the 'Enterprise of the Indies'. Of course, Columbus never really knew what he himself meant by 'the Indies'. Sometimes it referred to India, sometimes China or Japan, sometimes the Spice Islands of the Moluccas in Indonesia. This sort of continental aggregation was both intentionally vague and didn't matter much, because all these places were synonymous with one thing: money.

Historians are agreed that 15th-century European goods amounted to little more than cloth and cabbage. You might trade your cabbage for some cloth, or vice versa, and if you were lucky you might come into some leather. But the 'Indies', whatever they were, promised things that were both lovely and expensive, like cloves, tea, silk, sugar, cinnamon and, of course, gold.

Neither Columbus nor Columbus's Brother came from money. They came from Genoa, where their father worked in one of the two European trades – cloth – as a weaver. But Columbus was ambitious and desperate to better his status, so while he successfully built a reputation as an able mariner, he also realised that his best chance of reaching the very top was to become a steadfast contrarian.

Argumentative, stubborn and wrong. None of these traits made Columbus any fun to hang out with, but they stood him in excellent stead for the years he and his brother spent trying to get the 'Enterprise of the Indies' off the ground.

And their first port of call was Portugal.*

THE PORTUGAL PITCH

Had there been an international Expo for mariners in the 15th century, Portugal would very likely have been the venue of choice.†

Stretched out along the Atlantic edge of the Iberian peninsula, Portugal's wide western window looked out across the vast ocean, making it perfectly alluring for any explorer bold enough to venture beyond the world described by Ptolemy. In the first half of the century one of King John I's younger sons finally took the country's maritime destiny by the scruff of the neck and literally pushed the boat out.

Henry the Navigator, as he appropriately became known, ushered in the Age of Discovery with his design for a new type of ship with a triangular-shaped sail. The 'caravel', as the vessel became known, was faster, more manoeuvrable and, because of its greater speed, capable of travelling greater distances before the biscuits went bad. This enabled Henry and his fellow Portuguese explorers to explore further down the west coast of Africa and venture out as far as the Azores in the Atlantic.

While the enormous bravery that such an adventure demands is easily glorified, it's important to remember that this was also the start of an enterprise synonymous with one of the darkest episodes in our history. The people and places the Europeans encountered

---

* We're hoping you read this in a laboured, almost-rhymes sort of way.
† It also might very well have been called the 'Explo Expo'.

were almost always left decidedly worse off through the exploitation of resources and the slave trade, both of which motivated the exploration of unknown lands. Portugal's growing wealth came at the direct expense and suffering of people, the legacy of which, even hundreds of years later, remains a deep and open wound. Columbus, as his more recent historical reassessment has reminded us, was pivotal in driving the expansion of the trade in slaves, although if Portugal had had its way, he never would have had the chance to do so.

Decades after Henry the Navigator navigated, Columbus arrived in Lisbon keen to write a new chapter in the story of Portuguese exploration. As it so happened, he'd married a woman who was the daughter of a Portuguese nobleman who'd sailed with Henry himself, and this gave the lowly Columbus his crucial 'in' with the new Portuguese king, John II.

Columbus and Columbus's Brother arrived at court in 1484, and like a medieval version of *Dragons' Den** – but where there was only one potential investor, King John II himself – nervously began their pitch.

As Columbus's Brother struggled to set up the easel for his maps, Columbus addressed the Portuguese monarch.

'Hello, I'm Christoforo Colombo,' he began. 'And this is my brother . . .'

He held out his arm to indicate it was his brother's turn to speak. An easel leg gave way as it collapsed to the floor, sending various scrolls rolling across the room.

'Oh, I'm Bartho—, sorry, er, I'm Columbus's Brother,' blubbered Columbus's Brother, standing to face the king, as if the easel hadn't been important.

Columbus impatiently clicked his fingers in the direction of the dispersed scrolls and Columbus's Brother scurried off to recover them and start again.

---

* Or *Shark Tank*, as it's called in America. Or *Who Wants to Invest in My Business?*, as it's known in Morocco.

'I know,' continued Columbus, 'that Your Majesty has long been keen on finding a sea route to the lucrative spices of the Indies, so that our ships may return laden with other people's stuff without having to accommodate a difficult overland stretch to the Port of Suez. If only . . .'

'Go on,' offered the King.

'Nothing,' said Columbus. 'It could never be done. Anyway, I'm aware Your Majesty has been sponsoring voyages to discover a southern tip of Africa, from whence the Indies could be reached to the east. But my brother and I have a simpler, safer and speedier proposal . . .'

'I'm ditching the easel,' whispered Columbus's Brother, and he approached the king with a scroll tucked under one arm. Two armour-clad heavies quickly stepped in, aggressively crossing spears to block his path and causing Columbus's Brother to let out an embarrassing two-tone squeal, flushing his older brother with sibling shame.

John, reassured that this man posed no physical threat, gave a barely perceptible nod to the guards, who withdrew their spears with a sneer. Columbus's Brother now unfurled his masterpiece in front of the king.

'What is this, Columbus's Brother?' demanded John.

'This,' he began, 'is a true representation of the world. As you can see, it's far simpler to reach Asia by travelling to the *west*. I can assure you, it's much closer than others would have you believe; my own calculations suggest it to be little more than 4,000 miles. If Your Majesty could spare just three ships, a good captain could cross this ocean and return with gold and other bounties.'

'That captain is me,' boomed Columbus, hands on hips.

King John sat for a moment in silence. He knew he wasn't qualified to assess the map's merits, but he didn't want to project his ignorance to the court. So he simply said, 'Interesting.'

'Perhaps,' piped up a courtier, perceiving the king's dilemma, 'Your Majesty could convene a special council to assess the claims of Columbus's Brother and his novel map. If the council agrees with his measurements, Your Majesty will then be in a position to decide whether to invest. Call it due diligence, sire.'

'Yes! Due diligence. That sounds good, let's do that,' proclaimed the second John, before shouting, 'Next!'

Passing a glamorous-looking Moroccan lady with a vegan skin-care line, Columbus and Columbus's Brother scurried back towards the lift, where, after the doors had closed, Columbus put his brother in a headlock and berated him for being the only cartographer in Christendom who couldn't operate an easel. (Columbus's Brother tried to break free with a retaliatory wrist burn, but being the younger brother, was too weak to inflict any meaningful damage.)

Over the following weeks, the Council of Geographical Affairs met to adjudicate on Columbus's Brother's workings. Contained within its ranks was none other than Martin Behaim, future globemaker, who was given ample time to study the work of Columbus's Brother, before including an identical set of exaggerated Eurasian measurements on his own work.

We don't know how Martin personally judged the proposal at the time, but collectively the Council dealt the Columbus brothers the first in what would be a string of royal rejections, concluding that:

1. The trip would be too risky and too expensive.
2. Columbus was a 'visionary', but he was wrong about his distances and measurements.
3. The trip ran counter to Portugal's commitment to finding an eastern sea route to Asia by circumnavigating Africa.

And so, in a huff, Columbus left Portugal for Spain.

Columbus's Brother, however, stayed behind. He'd clearly left an impression on the king, because he found himself working for him, compiling a new world map based on the latest Portuguese discoveries. This gave Columbus's Brother invaluable knowledge of the expanding world, which would soon come to the direct aid of the person his life had always revolved around: Columbus.

AGAIN, IN SPAIN,
IT MAINLY GOES THE SAME

Columbus described his years of petitioning monarchs for patronage as his 'years of anguish', and, without his brother by his side,
the years in Spain would be some of his toughest. Its rulers at the
time were Ferdinand and Isabella, whose marriage had famously
united the historically antagonistic kingdoms of Aragon and Cas-
tille, setting Spain on a path to superpowerdom.

As it would happen, shortly after arriving in Spain, Columbus
managed to make friends with the notoriously pious Isabella's con-
fessor, and – while we're not saying he blackmailed Isabella with
an illicit knowledge of her darkest secrets – he did get an audience
with her in 1487 to put forward his proposal.

Isabella listened, and it wasn't a 'no'. She liked the idea of
'setting up in Asia' so she could do the Lord's work of converting
everyone there to Christianity. But there were two problems: first,
she was busy trying to finish the job of uniting Spain by defeating
the Moors in the southern emirate of Granada, which was using
up most of the royal bandwidth, and second, she once again set up
a council to review the proposal, who, like the Portuguese before
them, decided Columbus's measurements were wildly off (they
were right) and the journey was pointless (they were wrong).

Columbus continued to petition Isabella, but he simply couldn't
get the thing over the line. Spain wasn't a big seafaring nation like
Portugal was at the time, and the money just wasn't there. Then, in
1488, there was a glimmer of hope – not from Spain, but back over
in Portugal, where Columbus's Brother still had the ear of King John
II. John had been dumping money into efforts to find an eastern
route to Asia and was getting a bit fed up with the endeavour. His
latest expedition, led by a man called Bartolomeu Dias, had failed to
return when expected and was now presumed to have failed. And so
John wrote to Columbus, describing him as 'our dear friend', saying
he would be willing to give the western route a punt after all.

Unfortunately for Columbus, he wasn't able to act on the letter in time, because in December 1488 Bartolomeu Dias *did* return to Portugal and, would you believe it, he *hadn't* failed. He'd only gone and found what he called the *Cabo da Boa Esperança* – the Cape of Good Hope – or, the bottom of Africa.

In short, Portugal had finally navigated a sea route to Asia. Over a thousand years earlier, Ptolemy had described the Indian Ocean as a closed lake, but Portuguese perseverance and blind faith that he might be wrong had paid off. They'd sailed to the Southern Ocean, opened up a lucrative trade route to the east and massively upgraded their world map.

This, for Columbus, was terrible news, and things went from bad to worse the following year when he lost his board and lodgings and ran out of money. Columbus, and his ideas, were firmly outcast.

But while there was Columbus's Brother, there was hope.

Portugal was, understandably, incredibly protective of its discoveries, meaning other nations knew very little about Dias's trip, least of all where the bottom of Africa actually was. But when Dias reported his findings back to King John, one of the people who happened to be present was Columbus's Brother, meaning he knew *all* the details – details he could perhaps manipulate when pitching to *other* European monarchs.

THE MARTELLUS MAP

In 1489 Columbus's Brother secretly copied the world map he'd produced for King John of Portugal, including the newly discovered Cape of Good Hope, and left Portugal to catch up with his brother. In taking this valuable Portuguese-owned geography intel* with him, he was risking his own head, but that's just the sort of younger brother Columbus's Brother was.

---

* Geographintel.

The siblings met up in Seville, where Christopher, barely even saying hello, grabbed Columbus's Brother by the collar and shouted, 'Money! I need money! What kind of brother are you if you can't give me money when I need it?' before giving him another dead arm. Columbus's Brother said he could get it, all of it, he just needed some time, and Christopher slowly released his grip. Columbus's Brother then hurried off to Italy, where he tried to get support from the Bank of St George in Genoa (he succeeded), and then raised more money by selling copies of his valuable Portuguese maps to anyone who could pay a high enough price.

Later that year – no doubt by pure chance – a man called Henricus Martellus produced a world map in Florence that just so happened to have an enormous amount of Portuguese intellectual property all over it. Both this map and another he produced a year later in 1490 are considered of huge historical importance as they're thought to be the last surviving pre-Columbian* world maps.

On the earlier, 1489 map, the edges are cramped due to the size of the sheets it was drawn on – and there's no Japan or Atlantic Ocean. The 1490 one, however, is more expansive.

The first thing you'll notice on both is that Eurasia, once again, is massive. Martellus didn't use lines of longitude and latitude, but he wrote occasional degree distances between certain points, and just so happened to mark Lisbon and the east coast of Eurasia as 230 degrees apart, almost identical to the distance that would be drawn on Behaim's globe in three years' time.

Coincidence? Or Columbscidence?

Most likely the latter, especially as this was the distance the Columbus brothers were espousing as they sought patronage from European monarchs. What's more, the 1490 Martellus map comprises sheets that have revealed tracings in the hand of none other than Columbus's Brother himself. His fingerprints are, quite literally, all over this hugely distorted map of the world.

---

* They're not technically 'pre-Columbian', because Columbus was still alive. We just mean that they don't show the Americas.

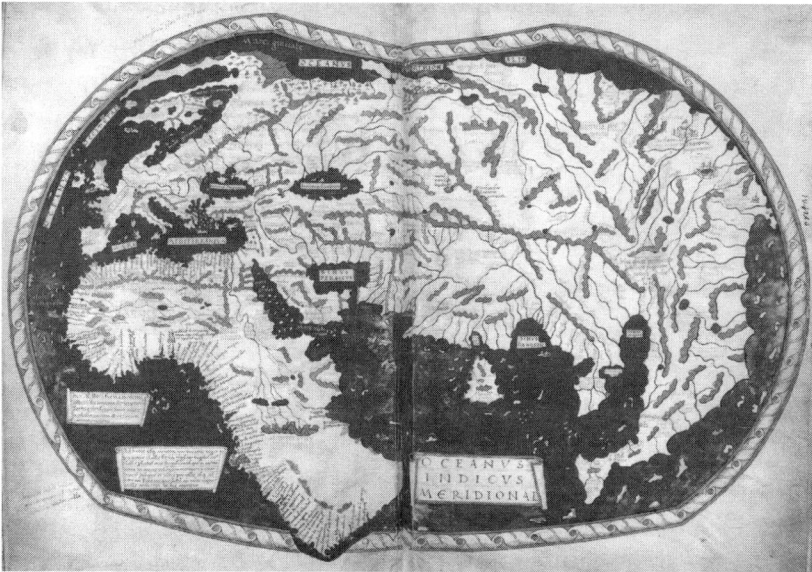

Above: Martellus's 1489 world map, British Library.
Below: Martellus's 1490 world map, Yale University.

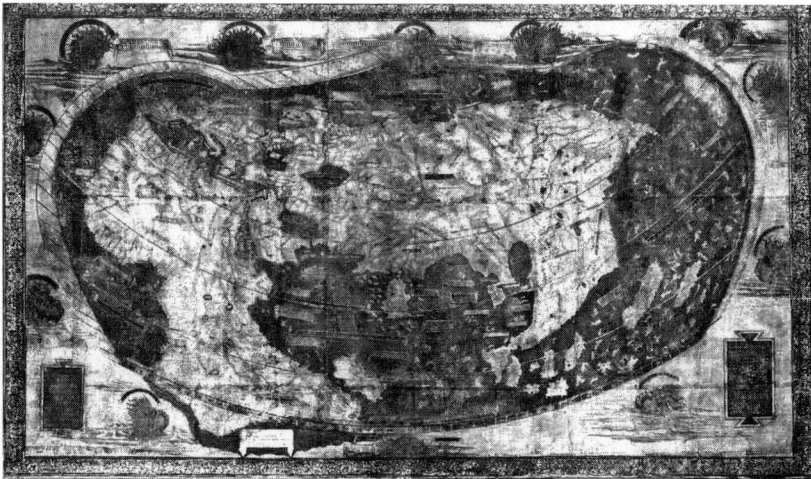

Which brings us onto the second thing of note. You'll see that, with Martellus's maps, something has gone badly wrong with Africa. Its southern tip is so ludicrously long, it is forced – on both – to break the frame. Discoverer Dias had marked the Cape of Good Hope at 34 degrees south of the equator, which just so happened to be bang on; A+ for his use of the astrolabe. On Martellus's map, however, it's an inexplicable 45 degrees south.

Was this Martellus's own assumption, or more of Columbus's Brother's doing? A clue may be found in some of Columbus's Brother's own personal writings found after his death, where he remarks:

> He [Dias] says that in this place [the Cape of Good Hope] he found by the astrolabe that he was 45 degrees below the equator. He has described this voyage and plotted it league by league on a marine chart in order to place it under the eyes of the most serene King himself [John II of Portugal]. I was present in all of this.

Columbus's Brother *was* present in all of this, and therefore would have *known* Dias located the Cape at 34 degrees, not 45. What he writes here, therefore, is a blatant lie. At 45 degrees, the journey east to Asia was made a whole 40 degrees less appealing (ten south, then ten north on the way out; then ten south and ten north again on the way back). One degree of latitude equates to 69 miles, so by adding a total of 40 degrees of sailing, Columbus's Brother had made the eastern route to Asia less appealing by about 2,760 miles.

The other standout distortion on the Martellus maps (also found on the Behaim Globe) is the shape of East Asia. You could be forgiven for thinking the Big Droopy Bit at the far east is the Malay peninsula, but it's not – that's actually the smaller droopy bit just west of the Big Droopy Bit. This non-existent Asian oddity, (variously named by historians as the 'Dragon's Tail' or the 'Tiger Leg', although we're sticking with 'Big Droopy Bit') can partially be explained as a hangover of Ptolemy, who, in encircling the Indian Ocean, had joined up the south of Africa with the east of Asia via a big stretch of circular land. Given Martellus was living in a world where maps were merely adjustments of Ptolemy, and the Indian Ocean had just been found to be open not closed, he could be forgiven for being unsure where to end the bottom of China.

But there's evidence that Columbus's Brother had, in fact, stretched this non-existent peninsula further south than would have been expected. We know that a key source for Columbus's Brother's own world maps was a Ptolemaic map by Lienhart Holle, whose East Asia only extended 15 degrees south of the equator before looping round to join up with Africa, but Martellus's ends up 28 degrees south, which is a lot further.

How might this second distortion have benefited Columbus? In two ways. First, because it indicated that even if you got round Africa heading east, there was a second inconvenience before you could reach the lucrative Spice Islands beyond China. And second, to any non-Portuguese monarchs, it suggested that it might well be worth sponsoring a western voyage, because the spoils of Japan, China and the Spice Islands would be protected from Portuguese traders, who'd presumably stop at India and then head back home.

Many historians have assumed that Columbus's ambitions were partly inspired by the Martellus maps' view of the world, but there's perhaps more evidence that it worked the other way round. Not only do the inexplicable distortions appear perfectly designed to facilitate Columbus's ambitions, they contain a body of evidence quite literally pointing to the hand of his brother.

But the crucial question remained: could Columbus's Brother's cunning cartography, showing such a heavily distorted world, convince a royal sponsor?

THE SPAIN CAMPAIGN
(BUT THIS TIME NOT IN VAIN)

Thanks to Columbus's Brother's timely financial assistance, by 1491 Columbus was back to his contrary, pompous self, ready to make one final appeal to Ferdinand and Isabella of Spain. We don't know what maps, if any, he had with him, but the Martellus Map gives us a pretty powerful clue as to how he might have described the world, and the ways, therefore, in which bad geography could have assisted him in persuading them to stump up the cash.

It just so happened that he arrived to meet them in a historic place at a historic moment, at the Spanish camp in Granada as the monarchs were preparing their final push to defeat the Muslim forces who'd occupied the southern Iberian territory for over two centuries. Columbus was there to witness their monumental victory and the raising of their flags, so he had both royals in the best possible mood for his proposal.

Columbus, the great persuader, pitched his heart out.

And they said, 'No, we're out.'

Apparently, some of their advisors had been won round on the geographical arguments, but now the sticking point was Columbus's own pride. He'd insisted on some excessive personal terms (albeit on a no-find, no-fee basis) including the title of Admiral of the Ocean Sea, and a 10 per cent commission on all ensuing trade from wherever he got to. He was sent packing for his hubris, riding off from Granada on his mule.

But, shortly after Columbus left, something interesting happened. The royal treasurer, having done a bit of a maths on a napkin, suddenly decided to intervene. He argued that the overall cost of the voyage was relatively low, and the rewards still potentially high, despite Columbus's demands. Remarkably, Ferdinand and Isabella, perhaps feeling the winds of divine good fortune on their backs, changed their minds. A court official rode off on horseback to give Columbus the good news.

The following year, in April 1492, Columbus sailed west on board the *Santa María* for an accidental meeting with history.

THE FATE OF THE YOUNGER

Columbus's Brother, meanwhile, remained a long way from any glory.

After his Italian escapades, illegally selling maps stuffed with Portuguese IP, he and Columbus set upon a divide-and-conquer strategy to seduce the richest European nations. While Columbus went to work on his relationship with Spain, Columbus's Brother was shipped off with his maps and his easel to court the court of Henry VII in Britain.

On the way he was captured by pirates, finally landing in Britain looking both sick and rather dishevelled. The famously miserly Henry VII dismissed Columbus's Brother and his easel, which not only continued to malfunction but had now begun to rot and corrode from months at sea.

Still, it found itself back on a ship and on its way to exhibit the same inaccurate maps to Charles VIII of France, also known as Charles 'the Affable'. But he wasn't, to the idea, and anyway Columbus's Brother soon received word that Columbus had not only succeeded in obtaining royal patronage in Spain, but had successfully returned, having (apparently) found Asia.

Columbus's Brother was overjoyed, and hurried back to Spain to congratulate him, and – who knows – perhaps even join him on his second voyage. When he got there, however, Columbus had already left on his second voyage. So Columbus's Brother could only stare wistfully across the ocean, happy in the knowledge that his brother was happy, as a single tear rolled down his cheek.

COLUMBUS'S BROTHER'S LEGACY

In the popular imagination, there isn't one. In terms of the course of history, however, there's a good chance it was massive. We can't know whether Columbus would have ever sailed to the Caribbean without Columbus's Brother, but there's one telling clue as to Columbus's Brother's significance in the venture.

Antonio Gallo – chancellor of the Bank of St George in Genoa – had met both the brothers and also lent them some cash. Gallo also happened to be Genoa's official chronicler at the time, and, after Columbus returned from his first voyage (wrongly claiming to have been in Japan when he'd actually been in Cuba), Gallo took a moment to record his impression of the brothers. He recalled that it was Columbus's Brother who'd instigated the 'Enterprise of the Indies' and the idea of sailing west to Asia. In his view, Columbus's Brother had entrusted the idea to his older brother Columbus because he was the better seaman.

Whether the idea was his or not, there's no doubt that Columbus's Brother's ability to manipulate the map of the world to make the western voyage seem more tempting might have helped tip the scales in Columbus's favour, especially as the final green light from Spain was given on such a knife edge.

Martellus's map and Behaim's globe are the lasting relics of Columbus's Brother's cartographical handiwork, showing the world just as he and Columbus argued it to be throughout the courts of Europe: an enormous Eurasia, a small Atlantic between Japan and western Europe, and – after Dias had found the Cape of Good Hope – an absurdly long extension to the bottom of Africa that broke the frame of Martellus's maps.

Columbus's Brother did, eventually, journey out to see the fruits of his enterprise, joining Columbus in the New World, and even travelling with him on his fourth and final voyage. By then things were going pretty badly, and not just for Columbus. The indigenous Carib people – many enslaved – were dying of novel European diseases in huge numbers and were on their way to being almost completely wiped out as a result of colonisation. Queen Isabella was shocked at reports of the treatment of the native people, whom she'd been hoping to bring to God through a peaceful mission. Even Columbus's own men had turned against him, as Spanish settlers and mariners lamented his tyrannical, increasingly unhinged rule.

But Columbus's Brother, it turns out, was cut from the same cloth as his brother.* Appointed governor of the island of Hispaniola (which today is divided between Haiti and the Dominican Republic), he was no kinder to his people, violently suppressing a rebellion in 1497 that arose in protest to his strict rule.

---

* Their father was a weaver, after all.

And it was here in Santo Domingo, on 12 August 1515, that Columbus's Brother died. In many ways, the more glorious adventure had been the one before the voyages and 'discoveries', a time when two lowly but ambitious siblings worked side by side, travelling across the royal courts of Europe, planning and plotting how they might persuade this king or that queen to invest in the world's most outlandish enterprise.

The reality of the voyages themselves is that they ultimately not only proved all their arguments about the world to be wrong, but also meant that the 'Columbus' name would subsequently become synonymous with slavery and exploitation.*

Today, Columbus's Brother's most obvious legacy can be found in the name of the tiny island named after him in the Caribbean. Nestled between Anguilla and St Kitts, the now French-owned island of Saint Barthélemy, or St Barts, was originally named by Columbus himself as an homage to his loyal sibling sidekick.

At least he didn't call it St Columbus's Brother.

---

* It also became synonymous with discovering North America, even though Columbus definitely didn't. And, in a second irony, Columbus didn't even get to name any of this New World after himself, because he never accepted that he'd even found a new continent. That honour fell to Amerigo Vespucci, an Italian explorer who, landing in Brazil in 1501, recognised that Columbus had quite definitely set foot in a 'New World'. Of course, this 'discovery' still needed the rubber stamp of a mapmaker, and when German Martin Waldseemüller became the first to write 'America' on his new 1507 world map, history was made.* (Waldseemüller based much of his map on the Martellus maps from 1489 and 1490, which had been based on Columbus's Brother's maps. Of course, Waldseemüller was able to make some vital corrections, although the bottom of Africa still breaks through the frame.)

* Waldseemüller only wrote the word 'America' tentatively across *South* America, but the map sold to the US Library of Congress in 2003 for a record $10 million dollars, which tells you something about the USA's determination to embrace both Amerigo and Columbus as part of North American history.

14

UNREADABLE STICK CHARTS (AND THE MAP THAT BLEW ITSELF UP)

*'Who lives in a pineapple under the sea? Absorbent and
yellow and porous is he.'*
Man who sings *SpongeBob SquarePants* theme tune

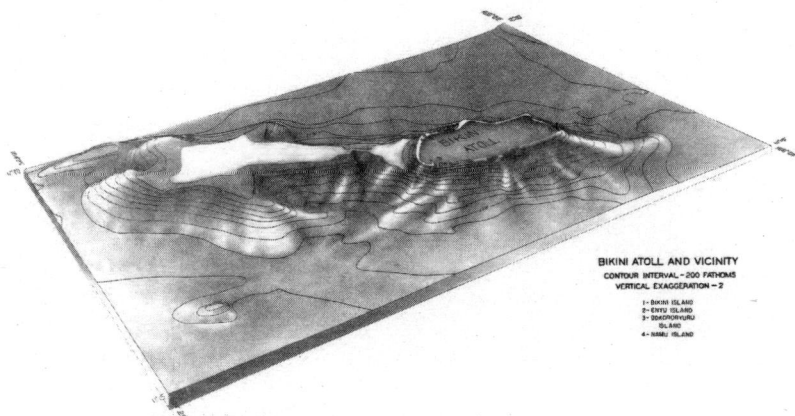

Bikini Atoll, 1956, US Department of the Interior.

This is the story of two maps of the same region, maps whose appearance and production couldn't be more different, that were each in their own way fascinatingly brilliant and yet that also each in their own way could leave you completely lost.

They're maps of a country with a combined total area of 70 square miles, and despite having nothing in common with one another beyond the broad territory they describe, their legacies are impossibly – and tragically – entwined.

In today's two-for-the-price-of-one chapter, we're in the Marshall Islands.

WHERE ARE THE MARSHALL ISLANDS?

Unless you live in the Marshall Islands, they're a long way away. Here's a map to prove it.

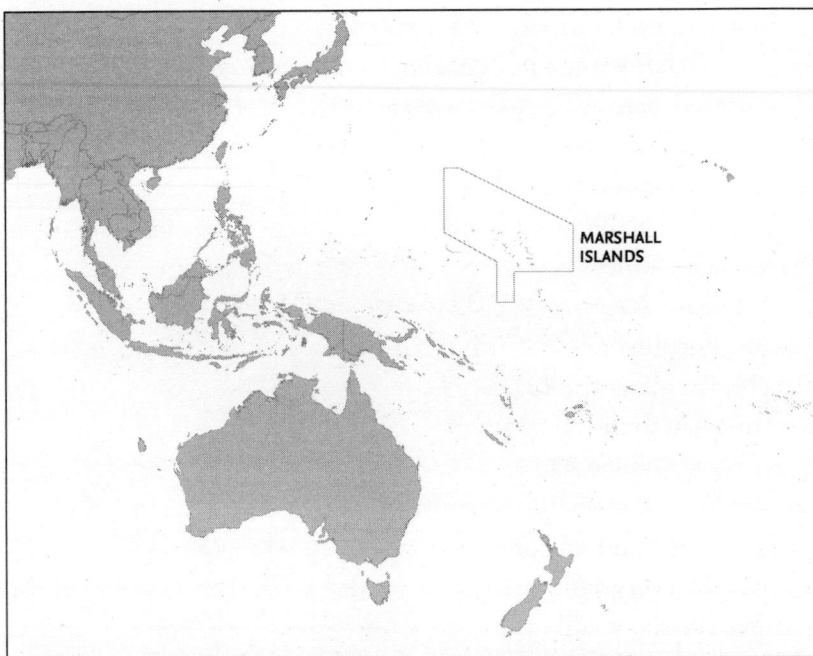

MARSHALL ISLANDS

First, some basic geo-historical facts. The Marshall Islands are made up of a pair of parallel archipelagos (island chains), and are home to more than 1,000 islets (mini-islands) concentrated around 29 atolls (circular reefs). They were named after British captain John Marshall (at best, the fourth Western explorer to spot them), but first colonised by the Germans. They were subsequently handed to the Japanese, then invaded and captured by the USA, before finally gaining full independence in 1979. Today, their biggest concern is rising sea levels. *Closes Wikipedia*.

Seen from above, the shape of each of the 29 atolls is akin to what a three-and-a-half-year-old might produce if tasked with drawing a perfect circle. In the centre of each atoll is a large lagoon, bordered by a series of mini-islands (islets) and narrow, broken sandbanks. The circular-ish shape of these atolls is no coincidence. The lagoons at their centre rest in the craters of long-extinct and eroded volcanic islands, around which the sediment from millions of years' worth of coral reefs have accrued to create land rising above the sea surface. Just. These are the islands that encircle the atolls, which, if you visit them today, are a paradise in appearance, all white sand beaches, coconut trees and blue waters.

Scattered across 750,000 square miles of Pacific Ocean, the actual land area of the Marshall Islands is exceeded by the surrounding water at a ratio of 1:10,000. The largest atoll is only 6.3 square miles small, so it would be impossible to get lost on any of these islands unless you were a minuscule non-human creature.

Travelling *between* the islands out in the Pacific is a different matter. For anyone who idly claims they've got a 'good sense of direction', this is the big league. You can brag all you like about the time you found a stationery shop in a medium-sized town even though your phone had run out of battery, but until you find yourself navigating without so much as a compass between tiny low-lying islands hundreds of miles apart in a featureless ocean so large it could fit every land continent inside it and still have room for another Russia, you'll only ever be a worthless orienteering novice.

The Marshallese, on the other hand, cracked it early doors, becoming probably the greatest seafarers who ever lived. Not only that, they also created ingenious and novel maps to pass their unrivalled knowledge of the seas on to the next generation, maps which are at once incredibly simplistic and impossibly complex, with rules that govern their production so arcane that, to this day, no Western scholar has been able to fully comprehend or interpret them.

MANY MAPMAKERS MOVE
TO THE MARSHALL ISLANDS

The origin story of the Marshall Islands is an explosive one, but the volcanic eruptions of their deep past pale in comparison with what was to come in the mid-20th century.

When the Americans captured the Marshall Islands from the Japanese towards the end of the Second World War, it was a deeply symbolic military victory. Less than two years earlier, the islands had been used as a staging post for Operation K, the second Japanese assault on Pearl Harbor. But beyond military symbolism, the Americans had little use for these tiny, remote, barely inhabited outposts. That was until 1946, when, all of a sudden, 'tiny, remote and barely inhabited' became precisely what they needed.

What had changed in the intervening two years was that the USSR had gone from German-fighting ally to staunch ideological enemy – their communist ways the opposite-est thing to the American Dream that anyone could imagine. Alarmingly, these communists were said to be developing nuclear weapons, which the Americans had already, but they now decided they needed bigger ones. And, putting the 'assured' into 'mutually assured destruction', these weapons needed testing.

President Harry S. Truman gave the Marshall Islands a call.

Chief Juda: Hello?
Truman: Helloooooooo! So good to hear your *voice*, chief . . . er . . . chief . . .
> *Sound of fingers clicking impatiently, muffled whispers still audible on the line.*
Truman (holding phone away from mouth): Name, name, quick, what is it? Give me the briefing file!
Chief Juda: Juda.
Truman: *Juda!* Juda the Dude-a . . . No? Never mind.

Aaaanyway, how are things? All good? American occupation treating you well so far?

Chief Juda: Er . . .

Truman: *Great!* You know we've always seen this as a completely 50–50 sort of bilateral relationship. In a way, you own *us* as much as we own *you*. And I want it to feel like that, in a way.

Chief Juda: What do you want?

Truman: Oh! Hahaha – no, no, nothing, just calling for a catch-up. You mustn't assume that every time I call it's because I *want* something. You know, we've got so much in common as *leaders*, you and I, because I think we both want what's best for the *world*.

Chief Juda: Uh-huh.

Truman: Oh! Haaaaang on a minute. You know what, actually, there *is* something, now I think about it. Literally just occurred to me. It's a small favour, but my sole concern is world peace, so I know you'll be in. We need to explode a little bit of ordnance somewhere.

Chief Juda: I see, well, I guess America is pretty big so . . . Oklahoma? Minnesota?

Truman: Hmm, yeah, interesting. I suppose . . . yeah. No, that is *one* thought, and I'll bear it in mind. Thanks for that. Yes. But, actually—

Chief Juda: You've already decided, haven't you?

Truman: There's a massive fleet on its way as we speak, yup.

Chief Juda: If you need to test a bit of TNT somewhere, it's not like I can say no, is it? We're not talking another Hiroshima . . .

Truman: . . .

Chief Juda: President? Surely you don't mean—

Truman: OK, well, nice talking, catch up later. Byeeee.
Dial tone.

Operation Crossroads was the name given to the nuclear tests that would be carried out in the Marshall Islands in the summer of 1946. Two bombs would be detonated at a carefully selected site called Bikini Atoll, remote even within the Marshall Islands, with the prevailing winds supposedly taking any fallout away from the rest of the archipelago.

Somewhat inconveniently for the Americans, Bikini was home to 167 people, but they were all warmly assured that their removal would be only temporary. Happily, according to the American military at the time, none of the Bikinians minded. In fact, according to one definitely not biased US Navy press report, they were 'delighted' with their eviction, as they'd been promised a new, better island where the coconuts grew bigger.*

The Bikinians were evicted in March 1946, the same month a fleet of American ships rolled into the atoll's serene waters carrying teams of geologists, oceanographers, biologists and environmental scientists – pencils hotly sharpened and ready to survey every inch of its islands in preparation for the enormous explosions. In the process, they'd make the most detailed maps that had ever been made of any atoll anywhere in the world.

A host of agencies and organisations descended on the fieldwork feeding frenzy: private research institutions, the Division of the Office of the Chief of Engineers, the Smithsonian, various university faculties, the Woods Hole Oceanographic Institution, the Fish and Wildlife Service, the US Geological Survey and many more.

Geological cores were drilled and analysed; wildlife was surveyed and indexed: birds, plankton, molluscs, land animals, reef fish, plankton – everything was to be studied, counted and charted, not least sea temperatures, currents, water exchange between lagoon and sea, and the size and shape of ocean sediment across

---

* As you probably guessed, the coconuts were not bigger on Rongerik Atoll, where the Bikinians were moved. Instead, two years after their resettlement, they had to be resettled again due to widespread starvation.

numerous transects of the lagoon. 'No stone was left unturned' was, in this case, not an analogy.

The unprecedented study was like a supersized version of a Year Nine geography trip to the coast to study longshore drift, only a) the measurements being taken felt like they might matter, and b) in a way that didn't feel entirely *positive*, depending on your personal level of antipathy towards communism.

The researchers produced new and detailed maps of the atoll against which future studies could be compared. Overall, as one report reported, the scientists agreed that 'Bikini atoll appears, on the whole, to be a healthy, flourishing structure,' a description that was doomed to age incredibly poorly.

One of the maps, in all its colourful glory, can be found in the colour section of this book. It's a contour map of Bikini Lagoon, with its 25 surrounding islands labelled in dark red. Today, there are only 22 and a half.

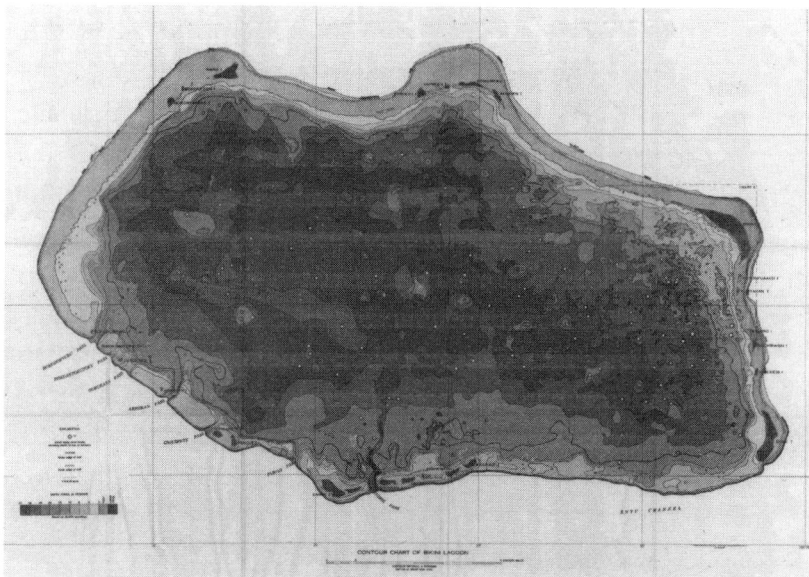

Contour map of Bikini Lagoon, with its 25 islands dotted around the lagoon, published by the United States Geological Survey, 1954.

CHOPPY WATERS

Stepping aboard the *Jitdam Kapeel* – a recently completed 36-foot traditional canoe with a sail and outrigger – Alson Kelen had the weight of history on his shoulders. It was June 2015, and he was attempting to sail from Majuro Atoll, capital of the Marshall Islands, to Aur Atoll, 84 miles to the north, without so much as a paper map, let alone GPS.

A hundred years earlier, there would have existed a number of Marshallese master navigators, known as *ri-meto* – or 'person of the sea' – who would have made the same journey with incomprehensible ease. Today, the *ri-meto* and their unparalleled navigational skills have all but died out, and Alson's journey marked a significant moment in a wider project to prevent them from doing so entirely.

As a longtime apprentice to one of the last skilled navigators (who has since passed away), Alson was untested on a journey such as this. Adding to the pressure, the trip had been funded by three Western scientists hoping to gain some insight into how the likes of Alson and his ancestors were able to find their way with such unerring accuracy with absolutely no naval tech whatsoever.

Instead, Alson had at his disposal a remarkable affinity with the feeling of the waves. He'd be sensing clues in the rock of the boat to determine their power, angle, shape and spacing, using a skill known as 'wave piloting'. This knowledge alone would, in theory, be enough for him to know where he was and which way to point the boat, even sailing through the night. The chances of success looked well and truly shot when, not long after departure, the wind picked up dramatically, and the waves became unnervingly rough and choppy.

Among the scientific team who'd be following in a motorised boat (*with* GPS – they were leaving nothing to chance) and who'd raised tens of thousands of dollars in research grants to organise the trip were an anthropologist, an oceanographer and a physicist who'd helped discover the Higgs boson and had now become keen to understand a second phenomenon currently invisible to science; namely, what it was that Marshallese navigators could detect in the waves that nobody else could. As if any further pressure were needed, the *New York Times* was writing a feature about wave piloting that would be centred on the story of this journey.

But how does wave piloting work? Sure, a boat rocks against a wave, and that wave might come from a known direction, allowing you to guess where you're pointing. But it doesn't tell you where you actually are, or how far you've sailed, drifted and been taken by a current since last Wednesday. Also, waves are notoriously affected by a host of changeable variables, including wind strength, wind direction and their interaction with the underlying currents. Using waves to navigate between tiny islands would appear to most people to be about as helpful as trying to get to work by following a butterfly.

But that's precisely the magic and mystery of wave piloting – a skill unique to the Marshall Islanders. The swells that cross the Marshall Islands often form a long way away, usually blown by the north-east trade winds from as far afield as California and Alaska or at other times of year approaching from distant storms originating in Antarctica or the warmer waters around Indonesia. But when these waves encounter a small Marshallese atoll, the swells bend around the islands, pushing off in slightly new directions. Other waves will bounce back off the small landmasses, pushing against the dominant direction of the swells and creating minuscule alterations to their feel up to tens of miles away.

It was these subtle distinctions – normally completely unnoticeable to anyone else – that Alson was aiming to pick out among the choppy waters as he sailed off into the night, hoping to encounter land in the morning. For the fretful scientists trying to learn something about the craft of wave piloting, the idea that this might happen in such poor conditions seemed nothing short of fantastical.

Inevitably, given we've included this story in this book, Alson Kelen confounded their expectations. The following morning, research boat in tow, he breezily sailed his way into Aur Atoll to a hero's welcome, as the Aur locals celebrated the impressive feat with a feast that the heavily seasick Western researchers mostly pushed around their plates with their forks. The long-term future of the *ri-meto*'s inherited cultural knowledge remains uncertain, but Alson's success was an important moment that showed traditional navigational knowledge was not yet lost to history.

The Marshallese ocean-faring prowess has been eulogised by Western writers for as long as the islands have been known, but the near-disappearance of the *ri-meto* and their knowledge was precipitated by a seismic moment in the islands' history, whose epicentre was the very atoll Alson had called home as a child: *Bikini*.

OPERATION CROSSROADS

After all the busy fieldwork in – and mapping of – Bikini Atoll had been completed in 1946, Operation Crossroads was ready to go.

And, unusually for something at once so risky and militarily sensitive, it was a media spectacle. A hundred and seventeen newspaper reporters were camped on nearby ship *The Appalachian* for the first explosion, codenamed Able. The event was filmed, described and broadcast around the world, presumably to scare the USSR back into its ideological box and show them precisely who was boss.

As part of the experiment, a number of old American warships, as well as captured German and Japanese vessels, were dotted around Bikini Lagoon to see how they would withstand the force of the explosion. Many of them had livestock on board, including pigs, goats and chipmunks.

At just after 9 a.m. on 1 July 1946, a B-29 bomber dropped a nuclear device with a yield of 29 kilotons. It exploded 500 feet above the surface of the lagoon, as a crackly voice on the radio somewhat pompously exclaimed, 'Listen, world – this is Crossroads.'

The explosion began with a genuinely blinding flash of white light – so it was lucky all the spectators were wearing their regulation dark-tinted goggles. The light soon dimmed enough to make the explosion itself visible, as a massive fireball morphed into a column of fire that appeared to suck itself into itself while rocketing high into the atmosphere, seemingly – from a distance – moving in slow motion, but presumably appearing much faster up-close.*

The Able bomb had missed its target by half a mile, destroying many of the scientific instruments placed to measure its impact, but it still did plenty of visible damage, sinking five ships and ruining many more. As for the animals that had been placed on the ships, the good news is that no people died.

Twenty-four days later the Americans dropped a second Crossroads bomb, codenamed Baker, this time designed to detonate beneath the surface of the lagoon. When it did, two million tonnes of water shot up into the air, turning into an eerie white spheroid that soon dissipated to reveal a huge mushroom cloud. Documentary footage of this famous explosion was used in *Dr Strangelove* and the 2014 version of *Godzilla*, though if you're only interested in the explosion bit, it's quicker to find it on YouTube. The waterborne shockwave destroyed many of the ships set in its path, and the amount of radioactivity spread across the atoll by the enormous wet nuclear mist ensured that the Bikinians weren't returning to their island any time soon.

Operation Crossroads has been described by a subsequent chairman of the Atomic Energy Commission as 'the world's first nuclear disaster', which feels more than a tad harsh on a certain pair of Japanese cities. But this was just the beginning of the Marshall Islands nuclear tests. What's more, behind the scenes, a team of nuclear physicists were busy inventing a new type of bomb that would alter the US-made maps of the atolls beyond recognition.

---

* Witness testimonies from this distance are scant.

President Truman's prefrontal cortex: OK, he's asleep.
Time to figure this one out away from all the political
grandstanding. Should America develop the hydrogen
bomb? Gee, this feels important.

We've got all the information we need – the
recommendation from the Atomic Energy Commission
report, chaired by Robert Oppenheimer, which quite clearly
recommends we should *not* do this. OK. That feels pretty
significant. Then there's the National Security Council, who
agreed with Robert.

Good, well, that's that, I think. America should not be a
world leader in building ever more destructive weapons that
could lead to the innocent deaths of millions of people. We
ought to convene with the Soviet Union as soon as possible
to agree on an international standard for all things nuclear
going forward, to protect the human race at all costs and
Do the Right Thing. Just scanning for any other rational
counter-arguments . . . No, nothing there. Good.

President Truman's amygdala: Ahem.

President Truman's prefrontal cortex: Oh Christ! Shouldn't
you be asleep?

President Truman's amygdala: Can't. I'm terrified.

President Truman's prefrontal cortex: Of what?

President Truman's amygdala: *EVERYTHING!!*
Communism, Stalin, political weakness, big Russian bombs
. . .

President Truman's prefrontal cortex: No, no, no, this is *not*
one for you. I'm figuring this one out alone. This decision
needs a rational prefrontal lobe approach, we're not liaising
with the limbic system.

President Truman's amygdala: So you want America to
be burnt to the ground, do you? That what you want? For
the Soviets to develop bigger bombs than us, then almost
definitely fire them because they hate our fabulous free way
of life? You want that to be our legacy, all because you did

NOTHING? The world isn't rational. Get with the times, PFC.

President Truman's prefrontal cortex: I suppose it is a terrifying thought – *wait*, I know this trick. You're trying to override me with the power of primal emotion. We must rise above our worst natures . . .

President Truman's amygdala: *Bombs! Fire! Death! Weakness! Criticism! Failure! The end of America!*

President Truman's prefrontal cortex: Stop it! Just let me *think* for a moment. I need to reason this out again. The Atomic Energy Commission recommended—

President Truman's amygdala: *Aaarrghhhhhhhhgghhghhgh Develop the Super Develop the Super Develop the Super!*

In January 1950, after months of deliberation, Harry S. Truman approved the development of the H-bomb, aka the 'Super'. The 1949 Atomic Energy Commission report he had ignored warned that, should the bomb get the go-ahead:

Many tests may be required.

But first, scientists had to figure out a way of forcing atoms together (fusion) rather than just splitting them (fission). Which happened to be quite complicated. Thankfully for the Marshall Islands, the H-bomb remained a hypothetical. For now.

A VERY DIFFERENT MAP

Most of us probably feel quite confident recognising a map and understanding what it's telling us. Typically, it involves a number of wibbly lines drawn from a bird's-eye view. Maps tend to exist either as pieces of paper or in the equivalent digital form. They usually try to represent relative distances accurately, and they're made with usability in mind – i.e. they're supposed to be accessible. Sometimes the sea will feature on the map, but the detail is almost always in the land; this makes perfect sense, as we almost exclusively live and go about our business on the land, and the sea is nothing more than a large expanse of inconvenient wetness to be traversed in the name of trade and travel. Some maps like to add in the odd darker-blue ocean trench to enhance the visual appeal, but more often than not it's just a monochrome, featureless blue.

You know all this, of course, but we're spelling it out here because, in the case of the Marshall Islands' most famous maps, none of the above applies.

This is an example of a Marshall Islands stick chart:

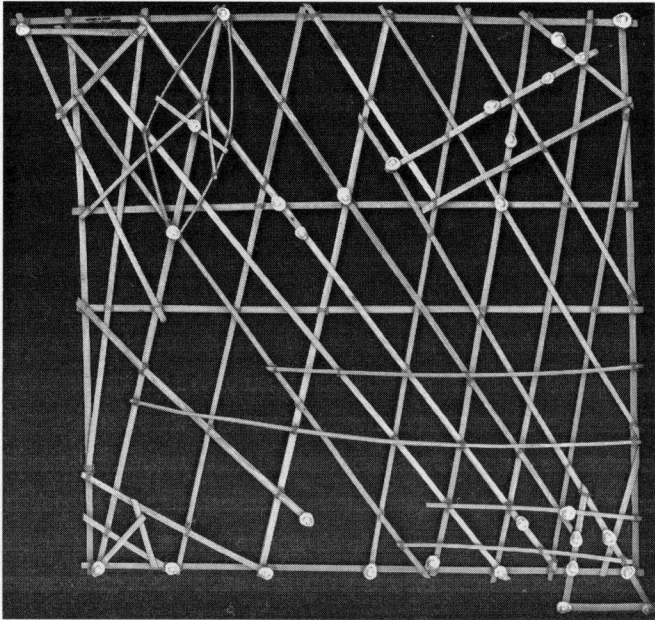

Rebbelib stick chart, 1920s, Library of Congress.

It shares the following similarities with maps everywhere else in the world:

- It's a bird's-eye view.
- It has lines.

That's about it.

The chart consists of a square stick frame with lots of criss-crossing sticks running top to bottom and side to side, all at various angles to the frame. Some are straight, some bendy, and they mostly but not always stretch entirely across the frame. The sticks are typically made from coconut fronds or pandanus root, tied together with a dried-grass fibre called sennit, with tiny cowry shells dotted about some of the lines, seemingly at random. They're not random, of course, because these small shells represent the atolls and islands that make up the Marshallese Republic's landmass.

That's how uninterested in the detail of the land stick charts are – the areas that people live on and call home are nothing more than a tiny shell. Instead, these are maps concerned with the dynamics of the ocean between the shells, and how best to get from shell to shell on a dark and cloudy night with nothing but the feel of the ocean waves to guide you.

Stick charts are the maps of the fabled *ri-meto* – the master navigators – a way of helping them learn, understand and pass on their unrivalled 'wave piloting' knowledge.

So, you may well ask, what do the lines actually mean?

Er . . . yes. Well. Quite. Um. See. We're not sure, really. Sorry.

Many a clever admiral has looked at these charts, looked at them again from a sideways angle, looked at them again from slightly further away, squinted a bit, held them right up to their eyes horizontally, and only ever wound up baffled. In fact, not even the majority of Marshallese can read one. Because stick charts do not speak a universal language, they were always person-alised, individually crafted by a *ri-meto* and made with their own particular set of rules, unknown to anyone but their creator.

Traditional Marshallese culture dictates that culturally impor-tant and valuable knowledge, such as navigational information, remains closely guarded, even between fellow islanders or fellow family members. The sharing of expertise would not only be heavily frowned upon by chiefs and other *ri-meto,* but dishing it out willy-nilly would only dilute one's own power and position as the holder of such valuable knowledge. As a result, their secrets would only be passed to their own navigating kin, or a chosen apprentice.

And you wouldn't want to be the Marshallese caught spilling the stick chart tea. The first Westerner thought to have been introduced to one of these maps was an American missionary called Luther Gulick in 1862.* He wrote an article praising the Marshallese's incredible understanding of the seas and their skill at navigating, also describing the 'rude maps by which they retain and impart knowledge'. He went on to explain that the person who'd so much as dared to show him a stick chart, without divulging any of its true secrets, had been threatened with death by the chiefs.

We do know what *sort* of thing stick charts show, collectively. We know for instance that the lines represent the directions of currents and swells. But unless you know which lines are which, it's impossible to know which lines are which. Sometimes a bendy stick means the direction of the current or wave bends, but sometimes the bend doesn't mean that. Sometimes sticks will be positioned to take the current into account, sometimes they won't. In short, these are maps that, should you try to repeat Alson Kelen's feat of sailing from island to island using one, will undoubtedly get you horribly and Pacifically lost.

Some of the charts are less mappy, more theoretically, and are known as *Mattang* stick charts. These were educational tools used to teach aspiring young *ri-meto* one of the core principles of wave piloting, and therefore Marshallese navigation as a whole: the ways in which swells are altered after hitting other islands in the archipelago. Understanding the subtle variations in wave strength and direction, and what that tells the sailor about their relative position to the other islands, is the key to Marshallese navigating prowess.

* He was also called Luther Gulick in every other year he was alive.

Mattang stick chart, pre-1941, British Library.

To help them learn this essential skill, apprentices would study their *Mattang* charts, before lying down blindfolded in the belly of a canoe in order to experience and learn the different wave movements at different locations. If they were unable to interpret the impact of the islands on the waves, they would never make it as a *ri-meto*.

It's hard to overstate the complexity of this skill. Not only did they have to understand the enormously varied patterns of the waves around the 29 atolls, how to recognise them from any number of different directions at an almost infinite variety of points around the islands and how these patterns change at different times of year, in different conditions or during an El Niño year; they also had to become masters of astrology, using the stars as wayfinding points and knowing which stars would be on the horizon of which islands at different times of night as the galaxy turned across the sky.

And even if they learnt all of that, it was worthless if they weren't any good at dead reckoning – estimating the distance travelled based on speed and drift since setting off – because they needed to have a feel for how far they'd gone to know which island's or islands' wave interferences they were feeling. Get it wrong, and it could be next stop, Chile. To top it all off, they couldn't even chat about any of this with their mates because of the intense secrecy surrounding the skills of each *ri–meto* and the knowledge contained in their individualised stick charts.

All of the apprentices' training would culminate in one incredibly stressful, potentially life-or-death test, overseen by their island's chiefs: a solo navigation between two distant islands. Miss your mark, and that was it. No second test. Find a new vocation (if, that is, they ever saw land again).

And, almost as if to show off, they did all this without a map to hand. Stick charts – the only material manifestations of the incredible knowledge of the *ri–meto* – were never actually taken on voyages, tradition dictating that they always remained on land. We're not sure whether this was because navigators didn't need them by this point, having committed their information to memory, or because they were too valuable to take in a canoe on choppy waters.

The skills of the Marshallese master navigators are a testament to what humans are truly capable of when it comes to finding our way. If they're able to cross hundreds of miles of landmark-less ocean with little more than the memory of a map made of sticks and the sensitivities of their inner ear to the feel of a wave, it truly puts into perspective our own inability to travel down a series of clearly signed roads between Leicester and Nottingham without slavishly following our sat navs.

So are all new advances in technology always a good thing? Well, the Marshallese have quite a strong view on that.

THE SAUSAGE OF DOOM

In early 1951 Françoise Ulam walked into the family living room and found her husband Stanisław staring out of the window. Later she recalled the odd expression on his face and how he'd turned to her and said, 'I have found a way to make it work . . . it will change history.'

Stanisław Ulam was a mathematician and member of the wartime Manhattan Project, who, when encountered by his wife that day, had just found a way to heat and compress thermonuclear fuel using the blast from a fission bomb to create a fusion reaction. With the help of his colleague Edward Teller, he had unlocked the next level in the Cold War arms race: the hydrogen bomb had arrived.

Since Operation Crossroads, two other nuclear-testing operations had been conducted in the Marshall Islands, totalling seven individual nuclear explosions, all of which had taken place on the recently evacuated Enewetak Atoll. Enewetak is the only atoll to the west of Bikini (and therefore the only one further from the majority of the islands to the south-east, the direction from which the prevailing wind comes). In fact, we can show you where it is by labelling the cowry shells on our stick chart from earlier:

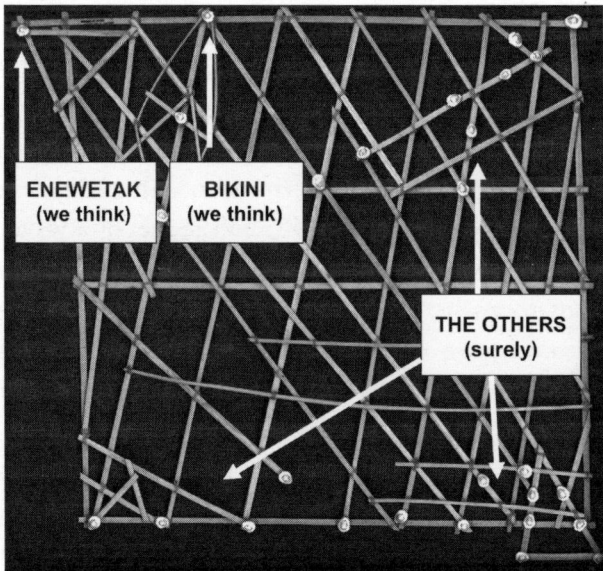

And it was here, on Enewetak in November 1952, that the first thermonuclear (aka 'hydrogen') device would be detonated in a test codenamed Ivy Mike.

And it *was* technically a device, not a bomb. The science was not, at this stage, ready to make a compact and 'deliverable' H-bomb that could be used if it were ever 'needed' in a war. So the reality was that the device in the Ivy Mike test was an impractical giant tank, housed in a two-storey building with a massive refrigerator attached to keep one of the key ingredients below –269°C, hardly suitable for a B-29 aeroplane. The tank's size and shape led to the device becoming affectionately known as 'the Sausage'.

And that was about the extent of its affectionateness. At 10.4 megatons, the explosion was so big that it created two entirely new elements never previously seen on earth. Its blast was 800 times the size of Hiroshima, and it instantly made redundant all of the detailed maps of Enewetak Atoll that had been carefully drawn up by the Americans in the preceding years, because the specific Enewetak island the Sausage had been assembled on (known as Elugelab) had now been completely vaporised. In its place was a crater 1.2 miles wide and 50 metres deep.

Apart from it working, the other good news for the Americans was that radioactive fallout from the Ivy Mike shot was relatively limited, as it had been taken up into the stratosphere where it would safely disseminate. In total, only 5 per cent of material from the explosion was accounted for in the aftermath, which, they presumed, boded well for all future H-bomb tests.

Now they needed to harness this power in a deliverable bomb. The boffins of doom got back to work, and two years later, in 1954, the 'Castle' tests were ready to go. This time, the H-Bomb was starting to shape up as something which could actually be *used* to kill millions of people in self-defence.

It was decided that, by this point, Enewetak Atoll had been dealt its fair share of nuclear obliteration, and it was time to go somewhere else. And so, nearby Bikini – still embarrassingly yet conveniently evacuated eight years on from the first Crossroads tests – got the nod.

THINGS GET OUT OF HAND

The next thermonuclear explosion in the Marshall Islands is the one that has been by far the most written about. And for good reason. Codenamed Castle Bravo, it was to be carried out in the north-west of the atoll, near the island of Nam, sometimes called Namu.

After the success of the Sausage, the device for the Castle Bravo test had also been given a little nickname, this time the 'Shrimp', ensuring that the Marshall Islands had now been served a veritable thermonuclear surf 'n' turf.

At 6.45 a.m. on 1 March 1954, the Shrimp was detonated. The Sausage had had a total yield of 10.4 megatons, while the Shrimp was estimated to yield about six, which was still absolutely enormous. But in the event it yielded 15 megatons.

According to people interviewed after the event, even with the dimming goggles on you could see the bones in your arms silhouetted against the glare of the blast. It remains the biggest bomb ever detonated in American history. The explosion instantly vaporised ten million tonnes of sand, coral and water, which then turned into a 100-mile-wide fallout cloud. But this time, something both unexpected and terrible happened: the material was not drawn up and dispersed in the stratosphere as had happened in the Ivy Mike test. And, instead of travelling west with the prevailing winds, away from the majority of the islands, the fallout spread east. Whether the unexpected trajectory of the fallout was a last-minute change in wind direction or simply the result of a poor understanding of the forces at play in the atmosphere has been long debated – what isn't debated is that highly radioactive fallout rained down on the inhabited nearby atolls of Rongelap, Ailinginae and Utrik, all of which needed to be hastily evacuated. But they weren't. It took over 48 hours to do so, despite the presence of many nearby American ships.

A bomb this size that behaved unexpectedly was always going to have consequences. First in its path was a Japanese (of all nations) fishing vessel with the double misnomer *The Lucky Dragon*, which had been fishing to the south-east of the explosion. All crew members suffered acute radiation poisoning; one died six months after the blast as a direct result of the illness, while many others died in subsequent years from cancers and complications that could reasonably be attributed to their exposure.

Many US military personnel also suffered serious skin lesions and burns as a result of being too close to the blast, whose yield had 'run away' so much further than expected. But nobody was more affected than the Marshallese themselves. In the military's hubris concerning the expected size of the explosion and direction of any fallout, no islanders had even been told of the planned detonation, let alone had been evacuated as a precaution. Children played in the radioactive snow, unaware of its terrifying chemical dangers. Although US officials at the time claimed that no islanders suffered from radiation poisoning, the burns, lesions and lost hair told a different story. In the years after, miscarriage rates rocketed among women, a third of Rongelapese (people from the nearest island) developed thyroid issues and 90 per cent of Rongelapese children developed thyroid tumours.

Adding insult to injury, the US then embarked on a dehumanising study of Marshallese 'subjects' for several decades and without their consent. Project 4.1, as it was known, included removing both healthy and unhealthy teeth from affected islanders, and forcing women to undress in front of them as part of the research into how people were affected by radiation poisoning.

By 1956 Rongelap was, according to Atomic Energy Commission director of health and safety, 'by far the most contaminated place in the world'. Inexplicably, his recommendation was then to send the Rongelapese people back to the island to study what happened as they continued to live there. He suggested justifying this by describing them as uncivilised and comparing the population to 'mice'.

We should add that the Marshallese are not the only victims of nuclear testing (America did also test a number of less powerful

weapons in Nevada), and the Americans were not the only perpe-
trators. Both France and the UK tested dozens of nuclear weapons
around the same time in French Polynesia, and the South Pacific
and Australia respectively, leaving their own legacy of devastation
that's a story in its own right. None of these three nations has yet
apologised.

But the scale of the explosions wrought on Bikini Atoll alone was
of an extraordinary magnitude. Despite Castle Bravo being billed as
the biggest nuclear testing disaster in US history, the military was
undeterred from its programme. Just 26 days later, in the exact same
location as Bravo, they detonated another bomb, in a test codenamed Romeo. Romeo had a predicted yield of four megatons and
once again 'ran away', this time to eleven megatons.

As you'd expect, the unrelenting tests in the Marshall Islands
have left behind a legacy of cultural trauma and destruction that's
plainly visible to this day.

CHANGING THE MAP

The atoll that had once been described by American scientists as
being a 'flourishing structure' looked very different when testing
was wrapped up in 1958. Over 12 years, 67 nuclear devices were
detonated in the Marshall Islands with a total combined force of
108 megatons, or 7,000 Hiroshima bombs. Twenty-three of them
were exploded on Bikini.

Today there are palm trees on some of its islands once again, but
unlike their pleasingly random spread on other atolls, the palm
trees on Bikini are found at eerily precise intervals, planted in neat
rows following the mass destruction that wiped them out during
the testing, with no thought to how oddly unnerving that would
look.

Other 'life' has returned in the form of some goofy cartoon
characters on a well-known Nickelodeon show who live in an
imaginary place under the sea called Bikini Bottom. The cast of

SpongeBob SquarePants could only have existed in a place as radi-
oactively volatile as the bottom of Bikini Lagoon. The writers have
confirmed some fans' suspicions that Sandy Cheeks, SpongeBob's
chipmunk friend, was a direct reference to the helpless chipmunks
stationed on many of the ships placed in the lagoon to test the
impact of the blast.

The bombs dropped on the Marshall Islands left their mark in
every sense imaginable. The Castle Bravo crater, for instance, is
still clearly visible from the air.

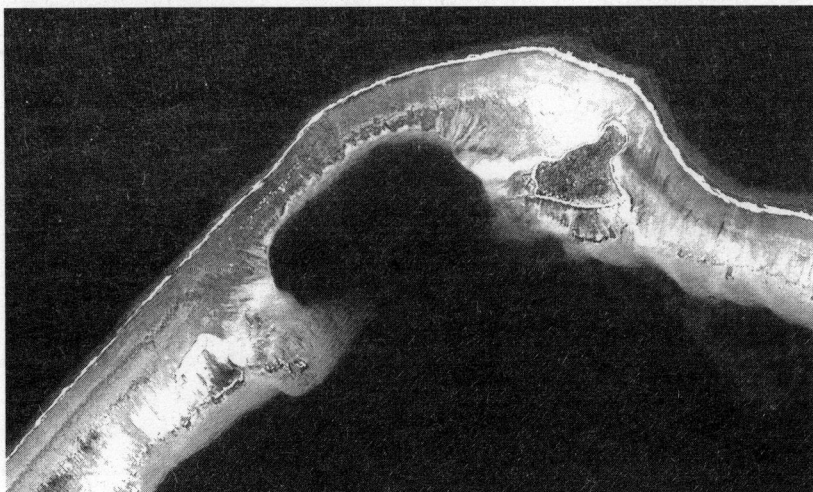

Previously, this corner of the atoll had been home to two islets
called Bokonijien and Aerokojlol, as well as the larger island of
Nam. Now, only half of Nam remains, and the islets met the same
fate as Elugelab Island on Enewetak Atoll after the Ivy Mike
explosion: vaporised. The loss of the islands on Bikini is reflected
on their flag, intentionally made to look similar to an American
Stars and Stripes, with 23 white stars representing the remaining
Bikini islands, and three black stars set well apart from these representing the lost islands of the nuclear tests.*

---

* A further two black stars – below the row of three – represent the two
Marshallese Islands most Bikinians live on today.

In 1946 each and every island on both Bikini and Enewetak atoll was not only named on the American maps of the atoll, but codenamed as well. Today, these names are much harder to find and identify. Google Maps only names three of them, and with no global superpower finding a Very Important Use for the Marshall Islands recently, it seems that nobody is particularly bothered with mapping them in the same sort of detail as in the mid-20th century.

To date, the Marshallese have received $600 million in compensation for the impacts of relocation and nuclear fallout, which many consider a pittance given the decades of environmental, human and cultural destruction wreaked on their country by the exploding of, on average, 1.6 Hiroshima bombs every day for 12 years.

Adding insult to injury, Bikini Atoll's name today is synonymous with neither scientific prowess, national security nor its location in the Pacific Islands. Bikini Atoll is most famously commemorated, in unbelievably bad taste, in a style of skimpy swimwear. The bikini swimsuit first hit the Cannes beaches after the 1946 Crossroads tests, when a clever marketing bod realised that linking racy swimwear with another 'explosive' moment on the far side of the world was bound to help it catch on as the latest fashion trend.

To this day, Bikinians remain dislocated from their homeland, as does the population of Rongelap Atoll. And when communities with valuable, fragile, inherited cultural knowledge are dispersed, the risk is always that the knowledge is lost forever.

THE LAST STICK CHARTS

Marshall Islands stick charts are about space as it's experienced, rather than as it actually appears from above, in that sense making them a bit like the map of the London Underground. The shells that represent the islands are not accurately placed to represent their positions relative to one another; they're placed to represent how journeys between them are experienced by the navigator, accounting for currents, wind direction and the strength of the swells.

The below diagram, for instance, shows the islands' true positions, adjusted for where they're found on this stick chart.

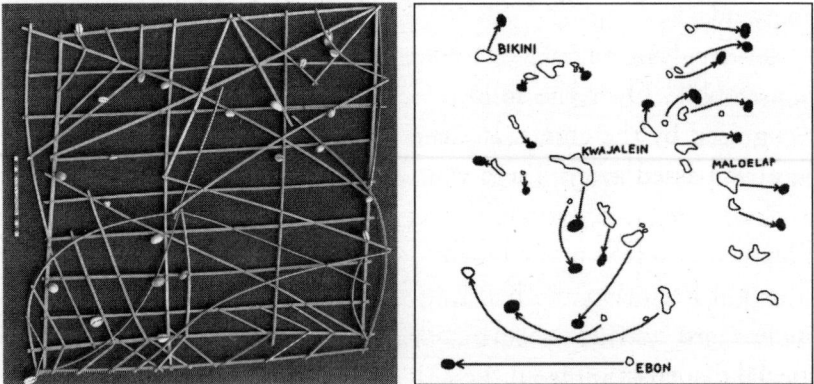

This positional 'roughness', combined with their unconventional twiggy canvas, means that stick charts have long been described as 'crude', 'rude' and 'primitive' by the anthropologists and geographers who first encountered and wrote about them. But of course, the information they contain is anything but. It's vastly sophisticated and specialised knowledge, a pinnacle of human achievement in wayfinding, and today on the brink of eradication. To be clear, wave piloting and stick charts were on the wane in the Marshall Islands before the 1950s, as the earlier arrival of Western explorers, their naval tech and different boats inevitably influenced the Marshallese's own approach. After all, why *not* use a compass just to check?

And, as ways of life changed and the number of *ri-meto* master navigators decreased, they concentrated their skills and teachings on one atoll: Rongelap. Here, in the early 20th century, new apprentices were still taught the ways of the master navigators of old and tested by chiefs to join the prestigious ranks of the *ri-meto*.

The system worked, and although the cultural inheritance inevitably diminished, it was sustained.

The Castle Bravo detonation of 1954 changed everything. Rongelap Atoll – just 94 miles east of Bikini – was among the first to be evacuated after the disastrous explosion. Many of the elders were reported to have died from their burns or radiation poisoning, and this last centre of *ri-meto* knowledge was scattered to the winds.

The resulting impact on the survival of the *ri-meto* skills was catastrophic. Over the following decades, no new *ri-metos* were recognised by the chiefs. In 2003, one of the last still thought to be alive passed away. There was one man left, though not quite a true *ri-meto*, a 55-year-old cargo-ship captain called Captain Korent Joel. He'd never taken his test because of the radioactive fallout at Rongelap, but had trained in the atoll as a boy before the nuclear test and learnt some of the vital skills. Korent asked for special dispensation from the chiefs to train his younger cousin, in a last bid to keep the craft alive.

That younger cousin was Alson Kelen, the wave pilot who successfully sailed from Majuro to Aur in the middle of the night and among the last Marshallese to possess the skill of wave piloting. Now an elder who runs a cultural preservation organisation called 'Canoes of the Marshall Islands', a non-profit teaching traditional Marshallese sailing skills and canoe building, Alson and all those who work with him represent a continued defiance of the cultural devastation of 67 nuclear tests in his native homeland.

As for stick charts, you can still find them all over the Marshall Islands, although they're rarely utilised in the same way as before. For all their cleverness, their fatal flaw was a frustratingly closely guarded instruction manual. When the fragility of this inherited knowledge was tested by tens of nuclear bombs, the manual was metaphorically shredded.

Today, they're celebrated as tourist curios sold from thatched shacks and Marshallese museum gift shops, as well as popping up on Marshallese stamps. Stick chart technology has been replaced by smartphones, the internet and Google Maps. The Marshallese have GPS, satellite phones and 250cc outboards on their boats. Life in the Marshall Islands is generally more prosperous, and most modern Marshallese are pleased about this.

But if their unique navigational knowledge were ever lost to history, there will be no getting it back. And that would be a terrible loss.

15

THE WORLD MAP
THAT WASN'T

10 August 1891

My darling Ida,

It is almost midnight, und I am riting from mein hotel room in Berne after a very suksessful evening at the International Geographical Conference.

Needless to say, my proposal has gone down a treat! You maybe are wondering why I have in English to you decided to write. Three reasons: First: to write in German would not be in the internationalist spirit of the evening. One must accept that, for ein great projekt such as mine to suckseed, one must enter into the spirit of globalism embodied by the Grate British Empire und the borderless wold that to build it is helping. Sekond, it gives me a good opportonitee to praktise my English letter-writing, of wich I will have to do a great deal lot of for the next few months. Und third, I am very, very drunk.

This evening's delegates have agreed with me. Wee need a uniform, topographic map of the wold in far more detail than those currently available. A standardised map covering every ocean und kontinent, that will enable strateforward comparisons to be made between different countries.

The world, smaller than it used to be it is. There is more international trade and travel ever than before. Children in skool are fascinated to be learn about far flung places. Und the missing piece in alle of this? A series of deetaled maps that will enable anyone: the explorer, the skolar, the merchant und the miner, to confidently understand the Geography of any place interested in they are.

This map, a compilation of meny sheets, will be of the bold skale of won to won million. On mo&t wold maps today, won inch is uzually around 800 miles, on mine – won inch will be 15.78 miles. Inkredable!

As a preeminent geomorphologi&t, I beleev that Geography is ein grate objectiv science. It is rational. It is absolute. Its impulses mu&t not be tamed by such arbitrary human con&tructs as national borders und intere&ts. The Himalayas, after all, are not contained by countries, deserts, rivers or Lower Triassic Bunter Sand&tone. Geography und its maps mu&t transcend nation-&tates, and for the common good of alle humanity function.

Below is the index map I hav being presented to them was, laying out the sheets that mu&t compiled be. Each qwodrangle on the below represents won map covering 4 degrees of latitude und 6 degrees of longitude.

Und now, mein Schatz, thanks to your humbel husband, the wold may soon have wot it so desperately needs: Ein international map of the entire wold at a scale of 1 to 1 million!

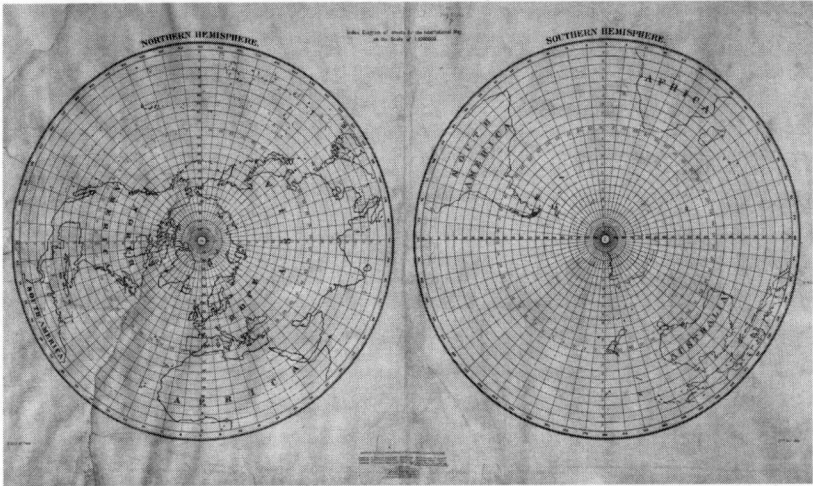

(Notice how my twin-hemispherikal projektion centred on the poles exaggerates the size of no kontinent. If the &trange bulging shape of Brazil is the price of an equal-area map, it is won I em willing to pay.)

I em so excited to get started immediately because, of corse, this is going to take qwite some time I think!

But, it will definitely be worth it. As they say in English: 'Rewarded shall be the man who does his foot cut off when hungry finds himself.'

Alle meine Liebe,

Albrecht

———•◆•———

11 August 1891

Dear Ida,

I em so very with hangover, und I fear I hav made a grate mistake.

I woke at 4am, anxious not only about my behaviour in the presence of the Russian delegate (apparently stroking his shapka is off limits!) but alzo – I fear I hav bitten off more than I can chuw. I counted all the littel sqwares in my index map from my last letter, und it turns owt we will need 2500 maps!

This is terrifying, no?!

Still, I hed a bath, und decided I must immediately rite to the governments of the big colonial powers – and make sure they are in agreement to start making maps of the territory they own, to a scale of 1:1 million as qwickly as possible.

But first, of corse, wee must agree a sistem of symbols und codes so that everyone can get to work on the map itself. The projekt will only sukseed if I can persuade all the mapmakers from all mapmaking nations to use exactly the same symbols, und pen thickness, und units of measurings, und level of the details, und fonts, zo that wen these maps all together are put they are the same looking.

Making them all to agree should the easy part of the job be.

Wish me luck!

Dein Herzchenflüsterer,

Albrecht

———•◆•———

January 1892

Cher Monsieur A. Penck,

Merci, or 'Many thanks' – as they say in English – for your letter from August last year. We apologise for the slow response. This letter has been sitting in our 'drafts' pile accidentally – and we only just noticed.

We think your International Map of the World is une très bonne idée. We do understand the reasons why – given the need for a standardised language – it should be in English, rather than French. Though honestly, Mr Penck, you didn't need to make the case over four pages – anyone would think we were sensitive about the issue.

No, we enter this with the open and forward-thinking vision you have proposed – this is about finding agreement and common ground, putting aside petty differences to find practical solutions to mapping our vast and spectacular shared planet.

And, having agreed to correspond in English, we feel sure that you will agree the Prime Meridian – the line from which all distances of longitude will be measured – must surely be Paris, with its noble and astronomic history as the foundational line of longitude.

We look forward to your positive response on this; needless to say it will be vital to our inclusion in the project, and therefore the guaranteed mapping of not only France, but also Algeria, Morocco, most of West Africa, Madagascar, Indochina, Cambodia, Martinique, Guadeloupe and the many other ocean islands that belong to us.

Meilleurs sentiments,

La France

P.S. Of course, you will be using the metric system. This is not a question.

———•◆•———

May 1894

Dear Mr Penck,

Greetings, old bean. Please do forgive the late reply, we just chanced to find a rather large pile of your grovelling letters in our mahogany in-tray marked 'spam'. We have now placed your letterhead in our database, however, so they should go to the right place from now on.

'International Map of the World'... Most interesting. We hear you've had a fair few rejections from the governments of plenty of countries who are neither willing nor able to stump up the time or money to offer a helping hand. Certainly, we can see why you require the big guns like us to get the thing moving, as it were.

Though we thought you should know that some of your arguments as to why we might help out were on the cusp of being obnoxiously obsequious. For instance, and I quote:

> 'A uniform map of the world would be at the same time a uniform map of the British Empire showing not only the actual territory under British authority, but also the sphere of British commercial activity, and would serve the varied purposes of administration, navigation and commerce.'

Is that your chat-up line to all the big colonial powers? Map tart!

Still, you're probably right, because it worked and we're in. It certainly would be rather helpful if one had a standard map of all the bits of the world one wanted to claim/exploit/rule/develop/help.

Just one potential spot of bother. It appears the French have your ear – why else would you write five pages on the history of the Paris Meridian, all of which is quite clearly obsolete given the agreement ten years ago at the International Meridian Conference in Washington that Greenwich should serve as a marker for zero degrees longitude.

Pas de chance, la France. That ship has sailed.

You'll have to go back to them on that, we're afraid. Unless, of course, you're happy for 25 per cent of the world to be missing from your world map, including not only Great Britain, but the Dominion of Canada, India, East Africa, the Royal Niger Company Territories, Burma, Guiana, Somaliland, Malaya, New Zealand, Australia, the colonies of southern Africa including Southern Rhodesia, North Western Rhodesia, Bechuanaland Protectorate and those three colonies we'll probably start calling South Africa in about 16 years' time, plus our many, many ocean islands.

Do let us know your thoughts.

Bestest,

Britain

P.S. We're assuming your suggestion about using the metric system was a sick joke.

———•◆•———

December 1908

My darling Ida,

I am somewhat exasperated. This is proving more difficult than I thought.

It's been seventeen years, and we haven't even agreed the symbol for a bus stop, let alone drawn any maps.

Maybe this was all a stupid idea, and I should just give up . . .

Deine Kuschelmaus,

Albi P

———•◆•———

December 1913

My darling Ida,

I'm writing to you late from my hotel room in Paris, and once again I am a bit tiddly, because I have some very exciting news. We have, finally, agreed on the standardised principles of the International World Map. And as it turns out – we're not even bothering with bus stops. PHEW!

Also, the project has even been given a cute nickname! Due to its iconic scale, it is to be known as 'The Millionth Map'.

You will also notice that the passage of time has seen my written English has vastly improved. Which you will be thankful for, I'm sure, as I know it would have become very annoying to read many more letters written in such a way.

Thank you, mein Liebling, for seeing me through all my moments of doubt these last 22 years – without all those motivational quotes you stuck up in the bathroom, it might never have occurred to me that 'the best view comes after the hardest climb'. But gosh, it really zings as a thought, does it not?

Tonight was the product of many years of hard work. Mercifully, Britain and France have agreed that the Meridian will run through Greenwich, while the measurements will all be in metric. Such a noble compromise. For a few years/decades I was worried that national interests might come before the great unifying force of scientific objectivity. But no! See the spirit of international co-operation in action!

To avoid bias, Latin characters will be used on all official sheets – though individual countries may of course publish sheets in their own script for national publication.

So far, we've had 35 countries come together to ratify the codes, all on board. The United States may be out (sadly they're a bit inward looking for now – it's just a phase), and anyway, Russia has signed up and they're much bigger. As long as Europe is on board – the fulcrum of international power for centuries to come – we should be fine.

Our project even has a new headquarters in Southampton, where none other than the British Ordnance Survey themselves will oversee the collection and compilation of the hundreds of one to a million map sheets that should begin flooding in imminently.

It appears we are on the cusp of a great age of international collaboration and enlightenment!

Now it's time to Draw. Some. Maps!

Dein Fröhlichkeiterreger,

Albi P

———— •◆• ————

August 1914

Dear Ida,
   Oh no. War.
   With our collaborating countries.
   Major problem, yes?
   Albrecht

——•◆•——

August 1914

Dear Ida,
   It was a rhetorical question. Of course it is a major problem – you
did not need to send me an entire thesis outlining all the reasons a
world war was a bad thing.
   Dein süßes Schmusekätzchen,
   Albrecht

——•◆•——

December 1915

Dear Mr Penck,
   Please destroy this letter after reading.
   Great care has been taken to get this letter all the way from
the new Millionth Map Central Bureau in Southampton to you in
Germany (where, thank goodness, you are now safely returned).
You will of course appreciate that we cannot be seen to have
corresponded with someone who has faced the dreadful
accusations you have been forced to withstand this past year.
   Who could have predicted this strange turn of events when
the Royal Geographical Society in London was awarding you its
prestigious Founder's Medal just last year? We were stunned to
read that Scotland Yard had arrested you shortly afterwards –
accused of being a spy! Perhaps that is the inevitable fate of a
German caught with a briefcase full of maps in England these days.
   But how miserable for you to have been put through the wringer
by the blasted tabloids. Typical of them to jump all over the story
as they did, and it is sad to see such jingoism thrust upon such

a renowned geomorphologist, not to mention one who has spearheaded the cartographic campaign for internationalism for so long. The irony!

Still, we are pleased to read that after a year in captivity – for which we, as fellow Brits, are deeply ashamed – you have finally been released to return home to your wife and children.

We do hope this experience has not changed your positive internationalist outlook in any way. In these precarious times, we may need your inspiring vision and leadership to help carry the Millionth Map forward.

As you know, many countries' governments have put the enterprise on the backburner. Some, because they are busy working on maps at a larger scale for military purposes, of course, but also one cannot help but feel the whole 'world war' thing has resulted in a poo-pooing of globalist ideals in favour of more national interests.

You and I know better, of course. Undoubtedly, we are on the right side of history.

You will be pleased to read we have overseen the production of 100 sheets of British territory and sent them off to the Southampton HQ. Hurrah!

We look forward to meeting up again next year, when no doubt this unpleasant war business will be firmly behind us, and we can all meet up for tea and sausages! The former: ours; the latter: yours – but only by a whisker ;-).

Sincerely,
Ordnance Survey (UK)

———•◆•———

NOVEMБЕЯ 1917
СОМЯДDЄ РЄЙСК,
СНДЙGЭ OF THЄ GЦДЯD OVЭЯ НЭЯЄ. ШЄ'ЯЕ OЦТ. SOЯЯY.
LЭNЇN

———•◆•———

June 1919

Dear Ida,

Today's news is the final straw.

As you know, I have struggled these last few years to reconcile my absurd treatment by the British with my steadfast belief that science should always transcend national borders. My time in prison has left me scarred and embittered, and I do not know if I shall ever fully recover.

I came to Great Britain – a nation I had once thought of as a home of rational thought and scientific objectivity – in the spirit of international cooperation. To be falsely shackled on ridiculous charges of espionage, with no foundation or evidence, has decimated my faith in the ability of countries to collaborate cartographically.

In prison, I realised I missed two things above all. You, my love, of course. But also Germany. Its land, its soil, its superior culture.

Yet now, the 'Allies' (we'll see how long that lasts) have signed this treaty in Versailles, crippling our great nation with all sorts of punitive measures that will stifle our economy and world-leading scientific research. And to think – they used the British-made Millionth Map sheets as part of the 'peace process' negotiations. The insult!

Clearly, other nations are idiotic, and you can see why I must doubt the entire notion of internationalism in which I had placed all my faith. There is only one country that can be truly trusted with leading the world forward on the basis of rational science, and that is Germany.

I have therefore concluded that Germany is Europe's natural leader, and should be allowed to spread its greatness across as vast an area as possible. It needs an empire, and lots and lots of space to help spread out forward-thinking ideas. In fact, as a qualified geomorphologist, I happen to know better than anyone that Polish soil is, in fact, German in character. I think that could be the basis for a political argument . . .

If the Allies and their Geographical Societies are such great friends, I will leave it to them to take on the burden of pleading with cocky upstart countries like America to be involved in the Millionth Map. And good luck to them.

Dein kuscheliges kleines Zuckerhörnchen,
Albrecht

———— •◆• ————

August 1919

Dear Ida,

Thank you for your letter – in which I couldn't help but detect a clear tone of concern.

No, I have not 'gone mad'. I am simply, as always, objectively and rationally deducing obvious scientific truths. 1) Germany is great. 2) Greatness must be spread. 3) Germany must be bigger.

One cannot argue with the logic.

Thank you for your latest quote: 'In times of difficulty, those brave enough to stay the course will be victors in the end.' Unfortunately, that one no longer resonates with me. I prefer: 'You can't go back and change the beginning, but you can start where you are and change the ending.'

In truth, I have no idea what that means, but I assume it's saying I'm doing the right thing.

As it happens, I do still believe in an international world map, although in my mind's eye it's beginning to look a little different to my first draft. What do you think?

Dein pünktlicher Kuschelzug nach Liebenthal,
Albrecht

———— •◆• ————

February 1926

Dear Mr Penck,

Long time no speak, old chap! Is something up?

Would be good to have you back on board to help keep this thing moving. Turns out it is an absolute bloody nightmare trying to marshal all these maps, countries, geographical societies and whatnot. Who knew!?

Still, thought I would give you an update in case I could pique your interest in helping out once more. We have 44 countries on board (for now) and have published 200 sheets! Good start, eh? Only 2,300 to go.

Sadly, only half those sheets are actually consistent with the codes and standards we all agreed back in 1913. In fact, just 13 of them have done it exactly as we first stipulated.

But what's the fun in a totally standardised map of the world, eh? Good to have a few quirks here and there. Keeps it interesting!

Sincerely,

Ordnance Survey

———•◆•———

April 1930

Dear Mr Albrecht Penck

We and our fellow African nations are getting a slight sense of déjà vu. It seems the mapping of our continent has been divided among a number of European powers – and nobody has consulted us about whether we want to be a part of this or indeed whether we might be able to assist.

Either way, it's rude. Apparently, you're responsible for this thing, and we demand an explanation.

Yours,

Sudan

———•◆•———

January 1939

Hey Albert,

Bet you didn't expect to hear from us! Got some good news for you, buddy – the American Geographical Society has almost finished mapping the entirety of South America at a scale of 1:1,000,000!

Take a look. Here's one of our maps with Chile and Argentina on it.

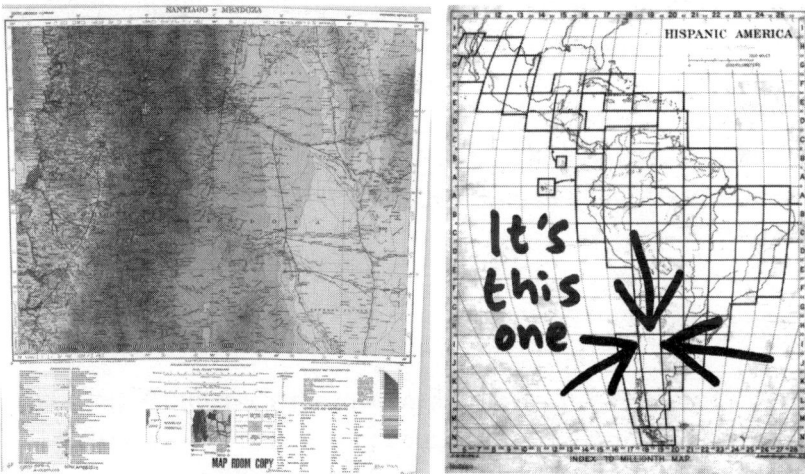

Pretty impressive, huh? Still, we're not going to bother with our own country. Problem is, we're still quite busy making detailed maps of all this new territory we've claimed – the USA's real big these days!

We kind of miss all your grovelling letters you used to send – it was nice to feel wanted. What happened?

Warmly,

The American Geographical Society (AGS)

September 1939

Dear Ida,

A great moment in history! It seems the German administration has understood that Polish soil is German soil, and gone to rightly claim it as theirs.

And while I will admit that Hitler is a bit 'out there' as a character (and yes, as an academic who has been closely watched by the Gestapo, you know I do not approve of their methods), the main thing is that Germany has the chance to expand its empire once more, to spread its enlightened ways among the more backwards regions of Europe.

I realise you have long since put away the motivational quotes, but I still often think of them: 'If you can dream it, you can do it' – well, it seems Germany has dreamt of a better future for all of Europe, and we are on the cusp of finally doing it too.

I imagine this will be a big headache for the Millionth Map, relieved it's no longer my problem!

Dein honigsüßes Marzipanherzchen,

Pencky poo

P.S. I know you're still upset with me over my new political work, but perhaps I could just come next door and chat instead of having to write you all these letters? Also, could you bring me a biscuit?

———•◆•———

March 1940

Dear Mr Penck,

In these dark days for our once closely allied nations, I send you an update merely in the globalist spirit in which your International Map of the World was first conceived.

To date, we have collected roughly 400 maps of the 2,500 needed to complete the project. Thankfully, they are still being produced even during this terrible war, albeit mainly to assist in the killing of others. Of course, it remains unclear whether any of them will end up being handed in to the Millionth Map project at all – such is the level of secrecy surrounding cartography these days.

Notwithstanding the conflict, it turns out this is a rather difficult project to manage. With so many national map agencies, geographical societies and governments to wrangle, it's something of a struggle to stay on top of what has and hasn't been done. Countries are in one moment, out the next. Maps being made, then not. If only there were some sort of International Prime Minister with an International Civil Service backed by a well-enforced legal authority to take charge of such a thing. Alas.

Still – we muddle on! And in good news, the Soviets are back on board! They say they're extremely committed to the maps being as accurate as possible. The nice man with the moustache gave me his word on that, and he seems like a trustworthy chap.

Don't be a stranger,
Ordnance Survey

P.S. Did you ever have any thoughts on who was going to map the 1,526 sheets which are just ocean? Nobody's really started them yet.

*

December 1940

Dear Mr Penck,

Bad news. Our Southampton office, home to the Millionth Map project, has been hit by a 250 kg High Explosive bomb, destroying a large chunk of the records. (We did mean to back them up but hadn't got round to it.)

War is VERY annoying.
Ordnance Survey

———•◆•———

December 1944

Hey Alfred,

I can proudly announce that America and many other countries are now completely committed to producing a Millionth Map of the whole world.

Only, not your one - sorry.

You may have noticed planes have become the next big thing, and would you believe it, we badly need to make aviation maps of the whole world as fast as possible.

You probably heard that earlier this year, 54 countries came together in Chicago for the International Civil Aviation Conference. We've agreed that immediate work is needed to produce and maintain a World Aeronautical Chart.

And would you believe, it turns out the best scale is 1:1,000,000! We started off using your International Map of the World sheets as the foundation for the maps, but it turns out they didn't show what we needed them to, so we've had to start again.

I have a feeling this is going to take almost all the time of all the main national and international cartography agencies and their mapmakers. Sorry.

Where does that leave things for the IMW? Do you think countries will want to produce two, slightly different, million-scale maps simultaneously? Might do, who knows!? Good luck, anyways.

Oh, and it looks like we're planning a bit of a fight back with the Allies this summer, watch this space

The AGS

———•◆•———

February 1945

My darling Ida,

I am now severely ill, as you can see, right here at my bedside, and fear I will not see spring, nor get to witness the great turnaround in our recent poor fortune in the war.

Damn the Americans, it was going so well. I had so hoped to see a return to the glory days of the Second Reich. With the guiding hand of a benevolent German Empire, scientific advancement would know no boundary – literally!

In truth, I fear this Third Reich bunch have forgotten their good Christian values, and perhaps that is why we are falling short in our endeavours.

I think now of the final quote you pinned up to the bathroom: 'Co-operation is always more powerful than competition.' I see now what you were trying to tell me; you wanted the old Albrecht back.

I am sorry if I disappointed you, but I believe it is the world that changed, not me.

Now, I hope my death will bring something positive. Namely, that we will stop receiving letters from all these irritating Geographical Societies who seem to think I'm still interested in receiving updates on the Millionth Map.

I stopped working on that long ago, and it is quite frankly bizarre that they haven't done the same.

Liebe von deinem pragmatischen kleinen Knuddelbär,

Albrecht

———•◆•———

October 1945

Dear Albrecht,

Quick update following on from all that nasty war business.

Many of the maps are out of date and require revision. Consistency also remains a big issue in relation to level of detail in respect of settlements, roads, railways and other communication routes.

Fancy coming back on board to help? Otherwise, we might just pack it in.

Ordnance Survey, UK

———•◆•———

May 1946

Dear Albrecht,

We would like to introduce ourselves. We are a new international organisation based in New York, and are very interested in picking up and taking over the Millionth Map project, as it chimes incredibly well with our values. Now the war is over, Internationalism is all the rage again, and we're at the forefront of that movement.

The Brits seemed happy for us to take this on, by the way – they're far more focused on bricks and mortar than maps these days. Fair enough, right?

I know you haven't been part of this for some time, but we'd love to invite you to see our new cartographic office in Manhattan. Maybe you could give us a press quote for the big Millionth Map relaunch?

There are quite a few people who think this is a waste of time ('cartographic wallpaper', according to famous geographer Arthur Robinson – Pah!), so would be good to restore the original vision somewhat.

Yours kindly,

The United Nations

———•◆•———

December 1950

Dear Albrecht,

Hellooooooo? Anyone theeeeeeerrrreeeeee????!

If you have a moment, the Egyptian government are complaining that your chosen relief colours (green for low-lying) are inappropriate for a desert country. Suppose it does seem strange that their map is mostly green while Canada is mostly orange.

Perhaps we should reconvene on the initial codes. Thoughts?

The UN

———•◆•———

January 1951

Dear United Nations,

My name is Fredrik, I am a bricklayer living at this address. Mr Penck died six years ago. Please stop writing to me, I am not interested in maps.

However, your unsolicited letters about this ludicrous world mapping project you seem to have undertaken have irritated me so much, I thought I'd offer you my unsolicited thoughts in return.

One can see how, when Mr Penck first proposed this idea decades ago, it ignited the imaginations of geographical societies around the world. Were they not still riding high on the age of discovery? The scramble for Africa? Westward expansion across America? All these new horizons and vast open lands to be discovered and described must have seemed dizzying, and the idea of somehow uniting the world under the good Christian values of morality and modernity, intoxicating.

The 20th century, of course, brought a new international reality. Two world wars, territorialism, mistrust, mass destruction. Were these the principles they had imagined for the new 'common humanity'?

Of course, I understand the impulse behind your United Nations schtick, and the attempt to corral countries back together as the world licks its wounds. And yet, it appears a new, larger conflict is on the horizon – and mistrust between the two great superpowers is deeper than ever. Do you think they're likely to work together on a world map, under the shadow of nuclear war?

Then there are the practicalities. Even if you were successful in persuading every country to agree to have its territory mapped in exactly the same way as everyone else's (a scale large enough to take ages, but too small to be of much use to anyone), what was your plan for keeping this thing up to date?

But what do I know? I am but a humble German bricklayer who spends his days helping to rebuild our broken country. All I'd say is, from an outside perspective, persisting with this idea appears to me to be nothing but Sisyphean madness.

Yours faithfully,
Fredrik

———•◆•———

March 1987

Dear Fredrik,

Thank you for your letter. Sorry for the 36-year delay in getting back to you, we couldn't find a stamp.

Just wanted to let you know that we have finally decided to officially terminate our pursuit of the International World Map.

The world certainly has changed since 1891, and you were right, the 20th century wasn't made for such a project.

In the end, we managed an admirable 1,000 sheets of the initial 2,500, but we decided not to spend the next century finishing it off.

Never mind.

Sincerely,

The United Nations

P.S. Chin up. We have every reason to believe that as the new millennium dawns, countries will only become more co-operative, ushering in a world of peace, equality and sustainability, where bridges are built, walls come down, poverty is eradicated and hope flourishes through shared prosperity, justice and harmony for all generations.

Another forty years or so should do it.

16

THE MAP THAT'S
NEVER WRONG

*'Any sufficiently advanced technology is
indistinguishable from magic.'*
Arthur C. Clarke

'Aaaaaaaaaaaaaaaaaaaarrrrghhhh!!!!!'
Also Arthur C. Clarke, when he first
turned on a hairdryer

In January 2013 65-year-old grandmother Sabine Moreau from Solre-sur-Sambre, Belgium, got in her car to pick her friend up from Brussels railway station 90 kilometres away, roughly an hour's drive. She punched the name of the station into her sat nav and trustingly followed the instructions. Two days later, she arrived in Zagreb, Croatia, after a journey of nearly 1,000 kilometres. She had stopped several times for petrol and for sleep, but somehow only realised that something had gone wrong when her sat nav announced, 'You have arrived at your destination.'

I hope Mrs Moreau wouldn't mind us describing this as 'her own stupid fault', her cross-continental adventure being probably the most extreme sat nav blunder in recorded history. Come to think of it, we wouldn't be surprised if Sabine had an ulterior motive for secretly driving to Zagreb, citing 'sat nav shenanigans' after a change of heart when her family found out where she was.

But the likelihood is you've heard a sat nav story like this before – possibly even in your own family – because Mrs Moreau's mishap is, sadly, indicative of a wider modern problem. It doesn't have to be a two-day diversion across five national borders for your day to be ruined by an overreliance on your sat nav.*

There's no doubt that digital maps and their near ubiquitous use have transformed our world and how people interact with it on a fundamental level. But more worryingly, they seem to have transformed us too. The vital skill of navigation that our species has evolved to learn and use over millions of years seems to have been taken not just out of our hands by new technology, but out of our brains too. Could it be that the best maps humanity has ever produced are simultaneously the worst maps for humanity?

A LONG, LONG TIME AGO

Ever since the very first map was published in ancient Babylon in the 9th century BC,† there have been people who complain that they 'can't read maps'.

This might be a hard concept for us map geeks‡ to get our heads around, but for a lot of people, making the mental conversion between a bird's-eye-view representation of the world and the world around them as they see it doesn't come intuitively. Maps, with their myriad distracting symbols, arbitrary interpretations and teeny tiny text, can be intimidating, not to mention tricky to fold and unfold.

And so there's always been a demand for a method or a technology that can be used as an alternative to maps. People trying to get

---

* By the way, for any Americans reading, we're going to refer to this technology throughout this chapter as 'sat nav'. Americans, who boringly call it 'GPS', are missing out. (Jay has similarly strident views on the word 'trousers'.)

† Both the words 'published' and 'map' are a bit generous here. It was a couple of circles scratched onto a clay tablet that several historians had to stare at for a few days before realising it was supposed to be a map.

‡ You've made it this far into the book; you're one of us now.

from A to B want to be guided on a specific route, with idiot-proof turn-by-turn directions for the lonely traveller, eliminating the likelihood of accidentally ending up at Z.

Unsatisfyingly, especially if you've been sent back in time to stop the sat nav being invented, there's no single date we can all agree on when the sat nav burst – fully invented – onto the scene. A series of false starts that sort of count and slow incremental improvements over time to various unrelated inventions in a frustratingly non-linear fashion place the date of the 'first sat nav' in either 1990, 1986, 1978, 1971, 1930, 1250 or even earlier, depending on what you think counts as a sat nav.

If we're going by the loose definition 'a technology to guide a traveller specifically from point A to point B without having to read a map', this, perhaps surprisingly, *pre-dates* maps.

The Romans used to navigate from one city to another not with maps but with step-by-step instructions known as *itineraria*. The reason we don't have any maps surviving from the Roman Empire isn't because they stored them in a box that was too leaky but because, as far as we know, they just didn't have any. (And who needs a map anyway when all your roads are dead straight?)

For many centuries the maps that did exist were seldom used for navigation. Even the most detailed maps had a habit of being slapdash with scale and paid little attention to roads. This was partly because the technology for consistent accurate scale wasn't available, and partly because there was no demand for it. Maps were, after all, primarily works of art, decorative items to adorn pub walls and monarchs' offices.

But this would start to change in the 13th century with the invention of the sat nav.

THE SAT NAV WAS INVENTED

Matthew Paris was a 13th-century monk who did his monking at St Albans Abbey in Hertfordshire. When he wasn't devoting his time to God, he was devoting it to cartography, producing a handful of various 'best guess' blobby maps of the shape of the British Isles that were objectively terrible but outstanding for the time, as they were some of the first to even attempt the feat.

It was shortly after producing these Britain blobs that Paris invented the sat nav. Being a monk, he was a keen promoter of the pastime of pilgrimage. But pilgrimage in the 13th century was, for anyone attempting it, a daunting prospect – it meant journeying across thousands of miles of signpostless foreign lands hoping not to lose your way. Being a cartographer, he came up with a clever way of making the task more straightforward.

Paris was aware of the Roman handwritten *itineraria* and realised he could take them a step further, turning these lists of place names into a sort of linear map known as an 'itinerary map', showing a series of pictures of places drawn along a set route. These maps are most notable for their most unusual layout. Instead of one enormous sheet covering a whole area, Paris produced a series of long, thin strips, each covering one specific journey from point A to point B, so that travellers could pilgrim their way from London to Jerusalem, reassuringly guided step by step the whole way there.

These strip maps were not to scale and contained no orientation, and the route itself was simply depicted as one long, straight line, with the itinerary of places dotted along it. The idea was the traveller and his horse could unfurl the scroll during the journey and, knowing what should be coming next, ask everyone they encountered to point in the direction of the next stop, be it Canterbury, Babylon, Constantinople or Magaluf.

While it was a vast improvement on the mapless alternative that preceded it, this method did, of course, have its drawbacks. For a start, they could only be used for a small number of very specific – pilgrim-based – journeys. They didn't allow for detours, or exploring, or going anywhere else. They weren't even ideal for the return journey back the way you came, unless you were happy to read upside down or make your horse walk backwards.

So, while Paris's maps were certainly unique, they became

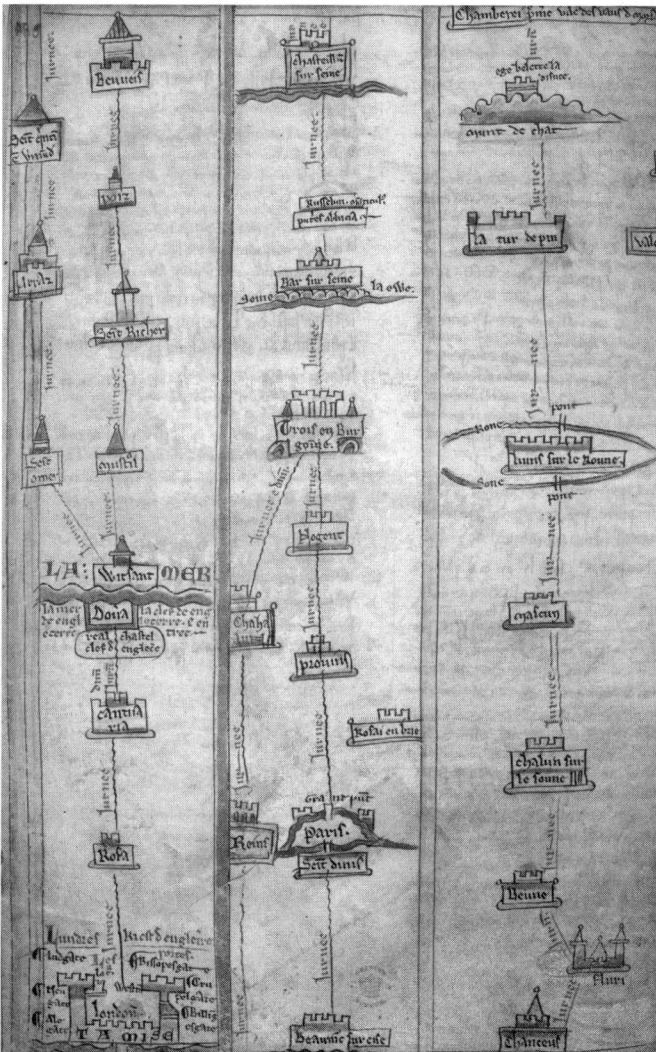

Matthew Paris's 'Itinerary Map', showing part of the route between London and Naples, from Book of Additions, 1250–59, British Library.

neither mainstream nor normal, especially as each one took ages to draw and produce. His concept of hyper-focused journey-specific wayfinding was a few centuries too far ahead of its time to catch on. But if Matthew had lived to 680 years old, he'd have been delighted to see his idea revived and improved upon.

THE SAT NAV WAS INVENTED

In the 20th century the advent of the motor car meant more people travelling further, more often, creating a more urgent need for a fail-safe navigation method.

So, in 1930 an invention came along called the Iter Avto, consisting of a long, thin map scroll, in a style very similar to Paris's, that was attached to a car's dashboard. When hooked up to the mileometer, the scroll would slowly rotate in perfect time with the car's movement, so it always showed the correct part of the map. It looked a bit like those bits of paper with musical notes on them that wind up on old-fashioned pianos.

But while the Iter Avto was certainly easier to use than the unfurl-it-yourself scrolls of the 13th century, it suffered from the same flaws, mainly that it was punishingly inflexible and required a different scroll for each journey, making its practical real-world use extremely limited. More importantly, with the high speeds that motor vehicles could now attain, it was very dangerous to use, as peering at tiny scrolling text while driving was less than advisable. On-board navigation could only really work safely in a moving car if the driver received spoken instructions while keeping their eyes firmly on the road, where they belonged. And this would take another 50 years.

THE SAT NAV WAS INVENTED

In 1971 a piece of experimental on-board audio-navigation technology was featured on an episode of the BBC's future-mispredicting show *Tomorrow's World*.[*]

A cassette tape gave pre-recorded audio directions synched up to the car's mileometer providing the driver with safe, clear

---

[*] *Tomorrow's World* was cancelled when we finally reached the future on 3 February 2003.

instructions at perfectly timed intervals in an experience remarkably close to the sat navs we're familiar with today. But despite the technology working a treat for the ten-minute demonstration on TV, this invention too had insurmountable flaws.

First of all, obviously, each specific journey required a different pre-recorded cassette on a fixed route, and an entirely different set of instructions for the journey back. So, with our realistic hats on, it's very difficult to think of literally any plausible practical uses for it. Except maybe these ones:

1. Hiring a car and driving from the airport to a specific hotel.
2. Being a *Tomorrow's World* presenter driving from BBC Television Centre to a random address.
3. . . . Nope. That's it.

And second of all, if a road was closed, or if you missed a turning, or even if you had to overtake a car with an unusually wide berth, the entire system immediately fell apart. And with the lack of any visual element for context, it was impossible to spot when anything had gone wrong. The cassette was unable even to tell you that you'd made a mistake, let alone how to fix it, and would continue cheerfully giving wrong – even potentially dangerous – instructions.

Oh, no, wait! We've thought of one.

4. A bus driver training on a new or unfamiliar route.

And adding insult to (sometimes actual) injury, the technology used the same cassette deck where your music would usually go, so if you were taking directions you had to take them in stony silence and couldn't listen to 'Chirpy Chirpy Cheep Cheep' by Middle of the Road, or indeed any other hits from early 1971.

Happily, things would improve massively a decade later when computers got involved.

THE SAT NAV WAS INVENTED

Converting paper maps for use on a computer is taken for granted nowadays, but it wasn't immediately obvious in the early days how to do it. There's another universe where reading a map on a computer requires typing the command 'turn page please' into a terminal. But thankfully, in the 1980s of *this* universe, Philips figured out how to store map data digitally, and how to display it.

The Philips Carin was a large, chunky computer monitor with an even chunkier full-size keyboard plonked next to the dashboard next to the driver's left knee (or right knee in left-hand drive countries where they drive on the right and sit on the left). Unlike the cassettes from a decade earlier, the compact discs in the Carin didn't contain a set of pre-recorded routes but stored maps digitally, enabling it to give directions from anywhere to anywhere the driver chose. It was even able to redirect you if you took a wrong turn. The monitor displayed a pixellated bird's-eye view of the surrounding streets that looked a bit like the 1981 arcade game *Frogger*, and was able to give audio instructions too. This was made possible by some impressive advances in robot voices that could simulate grumpy staccato human speech.

While it was a huge leap forward, the Carin, like all its predecessors, didn't *truly* know where it was. The driver had to manually program in their start position and hope that the computer, paying attention to the car's speedometer and its electronic compass, could keep track of where they'd ended up with simple dead reckoning. The slightest error in tyre pressure, and the whole thing would fly gradually off course, again making the system highly prone to potentially dangerous errors.

What all these experimental technologies had in common was that they were all, for one reason or another, *less* convenient than a traditional atlas, making them pointless. If on-board navigation was ever going to properly work and axe the atlas, the technology would have to, *somehow*, just always know exactly where it was, without the user having to constantly correct it. Was that even possible?

THE SAT NAV WAS INVENTED

It's time to *launch* into a discussion about SPACE! Blast off!!*

The US Department of Defense had begun working on a military project called NAVSTAR in the 1970s, and it was to change the course of navigation forever. After a lot of tinkering and asking the government for more money, in 1978 they flung several satellites into orbit above Planet Earth. With the right receiver equipment, a device could calculate the very, very, very precise distance away that these satellites were. This had the tremendously useful side-effect that, if the receiver could pick up a signal from at least three of those satellites, and as long as none of the satellites got bored and started floating somewhere else, with just a little bit of maths these receivers could calculate their *own* precise location anywhere on Planet Earth.

Here's a practical analogy to explain how it works:

Stand in a car park with a blindfold on and a blob of cotton wool shoved in one of your ears. Get one of your friends to stand at one end of the car park, screaming at the top of their voice while holding the note middle C. As you stumble around, the note will get louder and quieter, telling you how far you've drifted from your friend. Now bring a second friend along and get them to stand at the other end of the car park, screaming the note F sharp. Listening to the volume of these two notes, you can work out how far you are from both of them. This gives you a more accurate way of knowing where you are, but there is still some ambiguity. If both your friends appear to be screaming at the same volume, you could be standing at the main entrance or near the shopping trolleys.

---

* Editor: What are you doing?
Authors: We're quite tired now.

This is where your third friend comes in. Get them to stand in another position and growl a very low D. With these three bits of information, providing you remember exactly where you told your friends to stand, which friend is screaming which note and you're good enough at maths, you're able to perfectly triangulate your precise position with the use of just one ear.

That, but with satellites.

At first these receivers, which in the early days were about as heavy and cumbersome as a car, were for the exclusive use of the military.* The Global Positioning System, or 'GPS' to save time – as the military love to do – let troops, ships and aircraft pinpoint their exact locations anywhere on earth with the sort of military precision required of the military. It worked a treat, even in harsh weather or remote areas where traditional methods like maps or radio beacons fell short, making it possible to track military units and provide super-accurate coordinates for guided weapons. It had never been easier to blow things up.

But the practical uses of this technology for non-army people (and army people who'd quit the army) were obvious, so it was only a matter of time before the receivers got cheaper, smaller, and then cheaper again and smaller again, until eventually one of them turned up in someone's car.

THE SAT NAV WAS INVENTED

In 1990 military space technology and on-board map technology combined to produce the first commercially available on-board navigation system, though it was only available – for now – in one car, the Mazda Cosmo, an expensive sports car marketed at businessmen who were more concerned about showing off

---

* This is how most useful technology seems to start out. The 'army gets it first' policy applies to jet engines, microwaves, walkie-talkies, the internet, duct tape and baked beans (so they had something to put in the microwaves).

gadgets than arriving at meetings without getting lost. Other manufacturers targeting the same demographic quickly came up with their own versions, making it a common but expensive feature of executive cars throughout the 90s, much like walnut fascias. The real revolution came in the early 2000s, when sat nav technology got so cheap that it could be contained in a small plastic box that anyone could buy from Halfords. A little colour monitor with a touch screen could be shoved into the glovebox to be taken out when needed, attached to the windscreen with a little sucker that stayed on 80 per cent of the time and plugged into the cigarette lighter. Sat navs were now within the literal reach of practically every driver.

It's hard to remember now just how baffling and frankly creepy it was those short decades ago when most of us laid eyes on the technology for the first time. A small, cheap box with no moving parts somehow – magically – knew where it was. How was it so accurate? What was the catch? What *else* did it know? This early noughties iteration of the sat nav, which, by the way, came out at roughly the same time as the MP3 player, was arguably the beginning of the era where every new invention or improvement in technology would be a mysterious, unknowable process hidden inside a plastic casing, and might as well have been sorcery.

There was now a new sound effect to accompany all road trips. Along with engines revving, horns honking and seat belts clunking were the now familiar dulcet tones of instructions barked calmly from the sat nav – 'turn left, turn right, perform a U-turn when possible' etc. – sometimes voiced by Snoop Dogg or Stephen Fry. Traditional arguments between spouses about which route to take had now been replaced by similarly heated arguments over whether or not to obey the sat nav.

In-car navigation had become all but essential for lorry drivers, delivery drivers, private-hire drivers, bus drivers, ambulance drivers, firefighters, police officers, search and rescue teams, surveyors, civil engineers, construction workers, farmers, miners, road maintenance crews, your dad, tour guides, journalists and estate agents. Such was the ubiquity of the technology that safely operating a sat nav became part of the UK driving test.

There was no arguing that the world had changed for both the better and the safer . . .

Was there?

LESS THAN PERFECT

Like almost any new technology, the problems it solved were replaced by new, totally unforeseen ones. In much the same way that CD players eliminated the problem of magnetic tape being chewed up by the car stereo only to replace it with the much more annoying one of the music skipping and stuttering if you put only the slightest fingerprint on the disc,* sat navs too had their drawbacks.

Unlike paper atlases, sat navs depended on receiving a signal from satellites, and these sometimes failed. If your sat nav lost signal, it would forget where it was for a few seconds, or even a few minutes, showing a map of your car several meters away from the road. They had a habit of thinking your vehicle was facing the wrong way before you started driving, meaning lots of journeys had to be irritatingly re-routed within the first few seconds. They also relied on being able to see the sky to locate themselves, which meant that any time you entered a tunnel, the sat nav would act like you'd stopped dead in the middle of the motorway.

But aside from these mostly tolerable technical niggles, the concept of voice-led, computerised, mapless navigation had some pretty nasty flaws even when it was all working correctly. Sat navs often sent drivers on routes that weren't suitable for their vehicle. With the sat nav dutifully suggesting the 'shortest route', large lorries often wound up wedged under low bridges or stuck in tight

---

* Yet another thing *Tomorrow's World* mispredicted. They insisted you could smear jam all over a CD and it would still work perfectly.

corners on narrow country lanes. Landowners and local councils were forced to fight back, putting signs up everywhere, from entrances to properties, to junctions with narrow roads to the edges of cliffs begging drivers – those who noticed – to 'please ignore your sat nav'.

It seemed that the idiot-proof instructions of the sat nav allowed people's innate idiocy to go into overdrive. If you pop down to your local newspaper archive and ask them nicely to let you borrow the microfiche machine, a quick scan of articles throughout the noughties reveals hundreds of tiny news stories about people who blindly followed their sat nav and ended up in stupid places. Some of our favourites (including the absolute doozy from the start of this chapter, if indeed it *is* true) include:

- a taxi driver who'd been asked to drive from his customer's estate in Northamptonshire to Stamford Bridge, Chelsea FC's West London football ground, but instead drove 200 miles north to Stamford Bridge, a small village in north Yorkshire.
- a Swedish couple on holiday in Italy, trying to drive their hire car to the Mediterranean island of Capri, instead driving 400 kilometres north to a dull industrial town called Carpi.
- three Japanese tourists on holiday in Australia getting their car stuck in several feet of water after their sat nav told them to drive straight across the sea to reach Bribie Island.
- a 64-year-old granny in Leetonia, Ohio who couldn't see well in the dark, obeyed her sat nav's instructions to 'turn right', and did so onto a railway line, into the path of an oncoming train. (She got out of the car just in time.)

And those are just the silly ones. Sadly, there are plenty more stories from where these ones came that are more fatal than funny, where sat nav users became stuck in freezing mountains, got stranded in scorching deserts and drove off cliffs. The worrying paradox was that the more this technology tried to help, the more helpless the user became. And as the technology improved yet further, it would only get yet worse.

A development was just around the corner that would herald the unstoppable march of this technology permeating every facet of our lives, rendering us more helplessly at its mercy than ever. Sat navs were about to untangle themselves from the glovebox, leap out of our cars and into our brains.

THE SAT NAV IN YOUR POCKET

In a dazzlingly short period of time around the year 2009 and a half, everyone including your grandma upgraded their mobile phone from a device that could make calls, send texts and play *Snake* to a pocket-sized supercomputer with a full-colour touch-screen and access to the entire internet and, of course, a GPS receiver. The smartphone had arrived, and it changed everything.

As sales skyrocketed, sales of all the devices smartphones replaced plummeted, including alarm clocks, notebooks, diaries, Filofaxes, MiniDiscs, 10-megapixel digital cameras and, crucially, sat navs. And no organisation leapt upon this societal change, or drove it further, than one technology giant.

Google Maps has got at least one mention in almost every chapter of this book, and with good reason. In the 20 or so years since it launched, Google Maps has successfully become the unofficial-official default standardised map of the entire world.* And to discuss how that happened, the number of 'Google Maps' mentions is about to quinquagintuple . . .

When Apple released their first iPhone in 2007, Google Maps was one of the apps that came already installed. (This was in the days when Apple and Google were friends rather than the frenemies they are today.)

---

* Albrecht Penck (the one from the first half of the previous chapter) would have been very pleased.

Google threw everything into making their Maps product as high quality, free of mistakes and intuitive to use as it could possibly be, which made it a truly excellent sat nav. Navigating with Google Maps on your phone had several advantages over using the dedicated plastic box that lived in your car. First, your glovebox could now finally be a lot tidier, and second (more importantly), your phone had access to the internet. It didn't just know where the streets were – it knew *everything*. You could search for locations you didn't have the address for, locations you weren't exactly sure how to spell and even locations you didn't yet know that you wanted to visit (you could simply ask it for the nearest Mongolian restaurant).

But, of course, the biggest advantage of all was that through the magic of the internet, the maps were able to constantly update themselves. When a new road was built, or a street got converted from one-way to two-way, or vice versa, the map *knew*. Now the 'fastest route' really was the fastest, and the estimated time of arrival of '1.17 p.m.' really was 1.17 p.m.

The more people used them, the more accurate and useful Google's maps became. Users were encouraged to add their own data, such as where businesses were located. On a more passive level, navigating drivers could be prompted with a symbol for a road closure or accident with a button 'Still there?' And more passively still, the user – simply by existing – could gift Google live traffic data, even when the app wasn't being used. With so many millions of users leaving their phones switched on in their pockets throughout the day, Google was able to build a creepy stunningly accurate picture of where the traffic jams were.* With these jams getting worse in every city, Google's almost frighteningly accurate traffic predictions became far too tempting and invaluable *not* to use. Even people who knew their own city like the back of their hand found themselves tapping the screen at the start of every trip, including

---

* It was in the terms of service that you carefully read every page of before clicking 'accept' when you signed up.

their regular morning commute. Google Maps had transformed navigating by GPS from something that was sort of cheating to something you needed an extraordinarily good excuse to avoid.

And Google weren't just thinking of drivers. They honed their product to become an essential navigation tool for *all* travellers, including those walking, cycling or working out which train to get. While it would have been very unusual indeed to see a pedestrian walking around carrying a sat nav designed for a car, it became very common to see pedestrians staring down at their smartphone using Google Maps to find their way. But Google Maps went further still – beyond the realms of navigation.

Perhaps most cleverly (and profitably) of all, Google had the canny business sense to license their self-updating maps to third parties, powering other websites that use maps for all sorts of non-navigational purposes, such as by people who want to pay for their parking, track their run, order their Uber, agent their estates, Air their BnBs, and Go their Pokémons, smashing the virtual and physical worlds together, making Google's maps all the more ubiquitous and seemingly default.

So successfully did they blast their competition out of the water that in much the same way 'google' with a lowercase 'g' simply became a verb meaning to search online, 'Google Maps' too became a near-universal shorthand for any digital online map – much to the chagrin of their competitors. (See also tarmac, hoover, biro, aspirin, velcro and escalator.)

GOOGLE MAPS WILL GET YOU LOST

Happily, most people agreed that a win for Google was a win for humanity; a world in which the only way to get lost was to be irresponsible enough to run out of phone battery was, surely, only a good thing. Except . . . people were now ironically more lost than they'd *ever* been, precisely *because* of Google Maps' ubiquity and success.

In 2014 a study carried out by researchers at University College London suggested that our over-reliance on digital directions affected far more than simply how we navigate from A to B. They gathered 24 volunteers and told them to navigate their way around Soho in central London. The volunteers were split into two groups, with group A allowed to use their smartphones and group B told they had to keep their phones in their pockets. All volunteers had their brains hooked up to scanners on top of their heads. (This massive beeping equipment drew no attention whatsoever from passers-by; people in Soho are very used to seeing stupid hats.) Group B (who navigated without their phones) were shown to have significantly increased activity in two parts of the brain: the prefrontal cortex – which deals with decision-making, planning and reasoning; and the hippocampus – which is responsible for our spatial awareness, as well as forming and retrieving memories. As for group A, who were staring down at their screens, those parts of their brain were barely being activated at all. Their hippocampi and prefrontal cortices were, compared with the other group, effectively switched off during this task.

This chimes with the research mentioned way back in the introduction to this book about the changing hippocampus sizes of London's black cab drivers before and after retirement; the overwhelming evidence is the greater your reliance on a sat nav, the less you're working a crucial 'muscle' in the brain.

So, if we continue on this GPS-dependent trajectory, what does the future hold for our species?

A hundred years from now humans might become horrendously spatially unaware. People could become so dependent on technology that they're functionally helpless without it, unable to read a landscape, follow basic directions or even conceptualise where they are in the world if their devices fail. Imagine there's a sudden global outage after a solar flare knocks out all satellites or some meddling kids launch a cyberattack that switches off the entire grid. Whole populations could be stranded, not just physically but mentally, lacking the skills to adapt or problem-solve. Cities might plunge into chaos as people, unaccustomed to self-reliance, panic without their digital crutches, struggling to get by with nothing but their shrivelled hippocampi.

In the worst-case scenario, if computers take over navigation completely, human agency itself could atrophy. We'd be passengers in our own lives, vulnerable to whoever controls the tech. A malicious actor could manipulate maps to trap people, reroute resources or sow confusion, and no one would notice until it's too late. The very worst outcome? A species so infantilised and detached that we'd be at the mercy of failing systems or hostile takeovers, with no resilience to fight back.

Scarily, our species becoming specifically terrible at navigation is not even the *worst* worst-case scenario. Over-reliance on digital directions may already be having terrifying unseen consequences for the very bit of you that makes you *you*. Because the hippocampus, as well as being the bit of the brain that deals with spatial navigation, is also the home of memory.

Have you ever walked into a room and forgotten what you came in for? This is not a coincidence as it's known as 'the doorway effect'. Researchers at UCL have suggested that walking through doorways often causes us to forget things for the simple reason that they mark the threshold to a new space. Space and memory share the same cranial office for good reason; it turns out that space is an important dimension of *how* we form and preserve memories. It's the same reason most people find it easier to learn something that's been read from a printed page than from a scrolling screen: the strength of a memory is improved by the attribution of a three-dimensional geography.

Research into the impacts of GPS dependency is ongoing, but indications are that not only are we losing a crucial skill in itself, we may also be increasing our risk of developing Alzheimer's.

A 2024 study published in the *BMJ* found that taxi and ambulance drivers who'd mostly worked before the widespread introduction of GPS systems – and so spent their lives *thinking* about how to find their way around – had the lowest risk of Alzheimer's among 400 different occupations. The implicit suggestion was that the inverse might also be true – that *not* thinking about where we're going could increase the likelihood of Alzheimer's and subsequent dementia.

Widespread GPS use being fairly recent, of course, the final proof is still in the post, but with more and more studies seeming to confirm this trend, has the time come for us to take action?

'Hello, you're through to Google, how may I help you today?'

'Hello. This is Mark Cooper-Jones and Jay Foreman from Map Men, the authors of *This Way Up, When Maps Go Wrong (and Why It Matters)*. We're just calling 'cos we're concerned that the ubiquity of Google Maps has brought about an era where no one knows where they are anymore with troubling consequences, most notably the acceleration of an Alzheimer's epidemic.'

'OK, sorry to hear that. Can you let us know how we can make your experience better?'

'Yeah, could you delete Google Maps?'

'Yep, not a problem, let me juuuuust . . . *(sound of keyboard tapping)* . . . aaaand it's gone. So if anyone tries to open the app now it won't be there, and everyone should now be able to go back to navigating with a paper atlas just like in the olden days when everything was fine.'

'OK, that's great, thanks so much.'

'Anything else I can help you with today?'

'No, that's everything, thanks. Have a good day.'

'You too. Bye!'

'Bye!'

Unless the above happens, which we have to concede is unlikely (and unfair on everybody who uses Google Maps, including us), we must accept that digital maps are here to stay. But can we utilise their benefits and avoid becoming dumb slaves to their navigational prowess?

We think yes. There are a few simple things we can do to rekindle our brain's relationship with the physical world, keeping our hippocampi happy in the process. Happily, these don't have to involve buzzing around on a moped with a street atlas for four years, training to become a taxi driver.

One change is small but powerful: when navigating, switch to the overhead map view. This stops you being the centre of a universe that rotates around you and it keeps north at the top in a bird's-eye, non-tilted view. You'll still get the reassuringly safe turn-by-turn directions, but you'll also be given the chance to orient yourself, see the surrounding area and think about your location in relation to other places around you. This will keep the spatial part of your brain switched on and let you have a far better idea where you've ended up once you switch the engine off and get out of the car.

Or perhaps you're willing to go a step further and turn location services off altogether; on shorter drives, say (make all those road signs worth their enormous cost), or perhaps on a walk. You can still use your phone's excellently versatile and conveniently available maps, but add in the task of finding where you are and the whole thing becomes a fun daily puzzle, one where you might also find yourself spotting things you didn't know were there, because you're actually looking around you!

There are signs that we can be optimistic about the future. In recent years there has been a growing pushback against the ways in which digital technology has poked its way into every aspect of our lives. Even app developers themselves have cottoned on, implementing reminders for us to stop scrolling and take a break. More and more people are signing up to different forms of digital detox: club nights without phones, 'unplugged' cabin getaways, offline retreats. One survey suggested 86 per cent of people born between 1995 and 2009 (Gen Z) took steps to reduce the amount of time spent on social media in 2024.

A recent and particularly popular trend known as 'raw-dogging'* involves people wilfully banning themselves from looking at

---

* Don't ask.

any devices on long-haul flights, sitting perfectly still with no entertainment or any other form of external stimulation throughout the journey.*

We don't necessarily need to go so far as to both build our own traditional Pacific islands outrigger sailing boat and head out across the open ocean, but we could all think about the sorts of journeys we could make without the assistance of satellite navigation.

Who knows, it might even help stave off a public health crisis.

THE FUTURE OF MAPS

Assuming Google Maps and the like do not represent the last stop in the evolution of maps, we can't help but wonder what new technology will come along in the future that might make following arrows on a hand-sized device seem as archaic as flipping through the pages of an atlas. Augmented reality spectacles? Imperceptible pokes to the brain that make us feel like we somehow know our way to everywhere? A little mouse that lives in the brim of our futuristic hat tugging our hair left and right, and making the occasional wisecrack?† Self-driving pods and cars making us even more unaware that we've even moved at all, let alone knowing where we've moved to? The crumbling of civilisation around us negating the need, or ability, to leave the house again?

Will maps, as we currently recognise them, even exist at all? Especially given – as we've seen – they not only often go horribly wrong, but are by definition *always* flawed.

Happily, we think the answer is almost definitely yes, they will. There's nothing better for making sense of a vast and complex world than a map – and we'll always depend on them to help us make sense of it through the visual stories they tell. What's more,

---

* Apparently, staring at the in-flight map is allowed. We approve.
† Disney seem somewhat fixated on this idea, espousing it both in their 1953 biopic of Benjamin Franklin and the 2007 film *Ratatouille*.

there's an innate human desire to 'zoom out', to feel the power of perspective and appreciate where *we* are in relation to everything else. The best way to do this – other than getting on a plane or a rocket, both expensive – is with a map.

When Google Earth first came out in 2005, we rushed to download it onto our Compaq Presarios. And, with the entire rotating digital whole world at our fingertips for the very first time, what was the first thing we all did? Zoom in to find our own house. Our first impulse was to situate ourselves in the big wide world and feel connected to the rest of the planet.

That is the unique power of maps. Combine this with the fact that maps can be beautiful, intriguing and, from time to time, even useful, and there's a pretty strong case for them sticking around.

But what does concern us is that the joy of wrongness, in two different forms, is now at risk of being lost.

First, the sort of wrongness that's given us the stories we've told in the preceding chapters. The unstoppable spread of digital maps like Google Maps into every aspect of our lives means that, for many of us, there's now only one type of map that covers all the bases. Sure, for those of us who like a hike, you still need a decent topographical map as Google's not going to cut it across the fields or on the hill, but otherwise, why would you need anything else? With fewer different types of maps, and less need for original surveys as new mapmakers simply buy data from a limited handful of sources, there's an ever smaller number of idiosyncratic/unusual maps being made with different agendas and perspectives; maps that choose to show us something different from whoever's paying what to be shown on Google.

At the same time, there's a second type of wrongness being lost: the wrongness of getting lost in itself, which teaches you how to orient yourself and use the spatial part of your brain to find your way in the world. Thankfully, the loss of this second type of wrongness is more easily fixable than the first.

Grab a map, turn off GPS, leave the house, go somewhere. Could be somewhere you need to go, could be exercise, could be aimless. Turns out you don't even have to take a map of the city you're in; the chances are you won't end up in the mountains eating your own family.

Stop reading this book immediately. Put it down, and go and do the thing that we as a species have been doing – and been really good at doing – for millions of years. And have an absolutely lovely time while you're doing it.

Get lost.

ACKNOWLEDGEMENTS

MARK

There are many people we'd like to thank whose contributions have been vital for bringing this book together, without whom it would have been a quagmire of rank speculation and incoherence.

First, our editor Joel Simons, for his expert guidance and admirable patience in swimming through the gloop of our early drafts, and for giving feedback in such a manner as not to bruise fragile egos. Though, if we're nitpicking, it would have been great if you could have used Google Docs rather than insisting we reformat everything in Word every time. It was really fiddly and took ages 'cos the maps kept moving all over the place. Otherwise, spot-on editing.

Huge thanks also to the staggeringly capable team at Mudlark who have helped check, render, decorate and improve this book. Gaurika Kumar, Georgina Atsiaris and Mark Bolland for their exemplary editorial notes and assistance; Fiona Greenway for thorough and determined map-finding; Hetty Touquet and Chris Kwok for all their publicity and marketing ideas; Fionnuala Barrett, audiobook producer; Tom Dunstan and Dom Brennan in sales; Alan Cracknell in production; and art director Sean Garrehy.

Special mention also to Matt Burne, cover designer, for his patience with our notes over exactly how thick/what shape the arrow should be. We're really picky about arrows, sorry, but we couldn't be happier with the end result.

And to the true 'Map Men' of this book, the team who rendered and designed our maps for us – Robbie McGale, Kai Porteners and Jethro Lennox. Without you, both we and any readers would be well and truly lost.

To our agent, Anna Carmichael, at Abner Stein, for gently explaining how to actually write a book.

On this occasion we were *not* sponsored by Surfshark VPN, but we've given them a shout-out anyway for free, 'cos they've been very nice to us over the years.

There are a number of people we need to thank for their specific research contributions to various chapters. In no particular order, John Davies and Dr Alexander Kent, authors of *The Red Atlas*, for their map-finding, map-lending, fact-lending, fact-checking and time-giving, for seemingly nothing but this thanks – the least we can do is cross-promote; their book is wonderful, find it at redatlasbook.com.

Thanks also to Soviet cartography expert John Cruickshank and GDR secrecy expert Roland Lucht. To Dr Árpád Papp-Váry, son of Dr Árpád Papp-Váry, yet more thanks, for the invaluable translation of your father's work from Hungarian.

To former Google employees Elizabeth Laraki, Olga Khroustaleva and Janet Cheung, for sharing their story with us. You should all follow Elizabeth's substack: https://elizlaraki.substack.com/.

To mapmakers Alasdair Rae and Mike Hall for vital insight into how maps are made, and, confusingly, to Mike *Hills* for vital insight into how maps are made in newsrooms.

To Matthew Harris and Matt Gray (no relation) for both answering crucial questions about the history of TV, to Rob Watts for checking that a set of ludicrous German sign-offs made sense (blame him if they don't) and to Tim Byrne, who saved us from accidentally printing something offensive to French people.

And last, thanks to Woody Lewenstein for spending ages helping us with an entire bit of a chapter that we liked but decided not to use.

JAY

What he said.

BIBLIOGRAPHY

1 THE MAP THAT DELETED A COUNTRY

Patrick Evans, 'Ikea apologises after leaving New Zealand off a map', BBC News, 8 February 2019, https://www.bbc.co.uk/news/blogs-trending-47171599 (accessed October 2024)

'Ikea's map game is not on point', r/MapsWithoutNZ, reddit.com, 2019, https://www.reddit.com/media?url=https%3A%2F%2Fi.redd.it%2Fi7gswqwsq1f21.jpg (accessed October 2024)

IVAO XA & XU Oceanic Partnership, 'A Pilot's Complete Guide to North Atlantic Crossings', https://occ.ivao.aero/assets/docs/guides/APCGTNAC_EN.pdf (accessed October 2024)

Lauren McMah, 'Kiwi tourist detained as Kazakhstan officials say NZ is not a country', news.com.au, 5 December 2016, https://www.news.com.au/travel/travel-updates/travel-stories/kiwi-tourist-detained-as-kazakhstan-officials-say-n'z-is-not-a-country/news-story/38a2cb78c003859d071ff91f91783855 (accessed October 2024)

2 THE MAP THAT GUESSED

'1987 Territorial Growth of the United States Map', *National Geographic*, https://www.natgeomaps.com/hm-1987-territorial-growth-of-the-united-states (accessed October 2024)

Jonn Elledge, *A History of the World in 47 Borders*, Wildfire, 2024

'History', International Boundary Commission, https://www.internationalboundarycommission.org/en/about/history.php (accessed October 2024)

The Mitchell Map 1755–1782: An Irony of Empire, University of Southern Maine, http://oml01.doit.usm.maine.edu/special-map-exhibits/mitchell-map/introduction (accessed October 2024)

3 THE MAP OF THE WRONG CITY

Christopher Byrd, 'Guy Debord: The Life, Death and Afterlife of a Brilliant Crank', Hazlitt, 24 July 2014, https://hazlitt.net/feature/guy-debord-life-death-and-afterlife-brilliant-crank (accessed November 2024)

Guy Debord, 'Introduction to a Critique of Urban Geography', *Les Lèvres Nues*, No. 6, September 1955, https://chisineu.wordpress.com/wp-content/uploads/2012/09/biblioteca_introduction_debord.pdf (accessed November 2024)

Christine Donovan, 'Reinventing the City', *The Liminal Residency*, 29 September 2022, https://www.liminalresidency.co.uk/re-inventing-the-city/ (accessed November 2024)

Jaeneen, 'The Situationists: Psychogeographic Maps', *Medium*, 5 November 2017

On the Passage of a Few People through a Rather Brief Moment in Time: The Situationist International 1956–1972 (video documentary), 1989

Karen O'Rourke, 'Psychogeography: A Purposeful Drift Through the City', *The MIT Press Reader*, 16 July 2021, https://thereader.mitpress.mit.edu/psychogeography-a-purposeful-drift-through-the-city/ (accessed November 2024)

Sadie Plant, *The Most Radical Gesture: The Situationist International in a Postmodern Age*, Routledge, 1992

Ralph Rumney, *The Consul: Contributions to the History of the Situationist International and Its Time. Vol. 2: Conversations with Gerard Berreby with the Help of Giulio Minghini and Chantal Osterreicher*, Verso, 2002

Simon Sadler, 'The Naked City: Guy Debord and Asger Jorn', *Twentieth-Century Architecture, Part IV. Postwar Trends: Beginning Again, But Not at the Beginning*, 28 March 2017

Neil Scott, '5 Psychogeographical Experiments to See the City Anew',
    Crumble Magazine, No. 8, 17 September 2023
Paul Walsh, 'The Naked City', *Photowalk: Exploring Walking and
    Photography*, 9 July 2013, https://paulwalshphotographyblog.wordpress.
    com/2013/07/08/the-naked-city/ (accessed November 2024)

4 THE MAP THAT MADE UP MOUNTAINS

'African Association', *To the Mountains of the Moon: Mapping African
    Exploration, 1541–1880*, Princeton University, https://static-prod.lib.
    princeton.edu/visual_materials/maps/websites/africa/african-
    association/african-association.html
'At the Mountains of Kong' (video documentary), The Histocrat, 2021,
    https://www.youtube.com/watch?v=CI4la-iV4RY (accessed November
    2024)
Thomas J. Bassett and Philip W. Porter, '"From the Best Authorities":
    The Mountains of Kong in the Cartography of West Africa', *Journal of
    African History*, Vol. 32, Issue 3, November 1991, pp. 367–413
A. Adu Boahen, 'The African Association, 1788–1805', *Transactions of the
    Historical Society of Ghana*, Vol. 5, No. 1, 1961, pp. 43–64
Frank T. Kryza, *The Race for Timbuktu: In Search of Africa's City of Gold*,
    HarperCollins, 2006
Clements R. Markham, *Major James Rennell and the Rise of Modern English
    Geography*, Macmillan, 1895
Mungo Park, *Travels into the Interior of Africa* (1799), Eland, 2003

5 THE FUZZY MAP

'Belmont Transmitter and Yorkshire Television (Amalgamation)', *Hansard*,
    24 July 1970, https://hansard.parliament.uk/commons/1970-07-24/
    debates/962c17b0-a003-4e3b-8178-87a35831abc6/
    BelmontTransmitterAndYorkshireTelevision (accessed June 2025)

Simon Cherry, *ITV: The People's Channel: ITV 50*, Reynolds & Hearn, 2005

'Did Bilsdale Ever Transmit Yorkshire Television?', Vintage Electronics Blog and Forum, 2017–18, https://www.radios-tv.co.uk/community/colour-television/did-bilsdale-ever-transmit-yorkshire-television/

Matthew Harris, *ITV in the Face* (video documentary series) Bob the Fish Productions, 2019, https://www.youtube.com/playlist?list=PLO4G8mt6pQUVctOTSRvKUys8Gkbq8c37T (accessed June 2025)

Sophie Kinsella, *Shopaholic to the Stars*, Bantam Press, 2014 (but then we gave this book back, 'cos it turned out it was no help at all)

'The ITA Transmitter Network', ITA, https://tx.retropia.co.uk/ (accessed June 2025)

'This Is Transdiffusion: The Independent Broadcasting Authority Since 1964', https://transdiffusion.org/ (accessed June 2025)

'UK Broadcast Transmission', https://tx.mb21.co.uk/ (accessed June 2025)

6 IS A GLOBE A MAP?

Dietrich-Wilhelm von Schlegel-Hauptmann, Jean-Étienne de La Zouch, Astrid-Ludovika Kjellström-Van Der Waals and Stanislaus-Leopoldowicz Przybyszewski-Zamoyski-Krasnodębski, 'Cartographic Dimensionality and Spheroidal Representational Epistemologies: A Hyperorthogonal Disquisition on Globular Topologies Versus Planar Cartographies in Geospatial Ontologies', Annals of Arcane Geodetic Semiotics, Vol. 52, No. 4 (2026), 1893–2147. doi:10.1093/1234567890.

7 THE FICTIONAL MAP THAT BECAME REAL, THEN FICTIONAL (AND THEN REAL AGAIN (BUT ONLY FOR A BIT))

Snejana Farberov, 'Mystery of a town wiped off the map by Google: How the fake Agloe, New York, conceived as a "copyright trap" by cartographers turned into a real hamlet before vanishing forever', *Daily Mail*, 29 March 2014, https://www.dailymail.co.uk/news/article-2592500/Mystery-town-wiped-map-Google-How-fake-Agloe-New-York-conceived-copyright-trap-cartographers-turned-real-hamlet-vanishing-forever.html (accessed June 2025)

Frank Jacobs, 'Agloe: How a Completely Made Up New York Town Became Real', Big Think, 12 February 2014, https://bigthink.com/strange-maps/643-agloe-the-paper-town-stronger-than-fiction/ (accessed June 2025)

Robert Krulwich, 'An Imaginary Town Becomes Real, Then Not. True Story', Krulwich Wonders, 18 March 2014, https://www.npr.org/sections/krulwich/2014/03/18/290236647/an-imaginary-town-becomes-real-then-not-true-story (accessed June 2025)

'The strange story of Agloe, NY', *Times Herald-Record*, 30 October 2016, https://eu.recordonline.com/story/lifestyle/845-life/2016/10/30/the-strange-story-agloe-ny/24603254007/ (accessed June 2025)

8 THE MAP IN A BOX

'Ban on putting Shetland in a box on maps comes into force', 4 October 2018, BBC News https://www.bbc.co.uk/news/uk-scotland-scotland-politics-45733111 (accessed March 2025)

'Scotland's most remote islands don't want to be in "inset maps" any more', The Conversation, 6 November 2018, https://theconversation.com/scotlands-most-remote-islands-dont-want-to-be-in-inset-maps-any-more-106139 (accessed June 2025)

9 THE PARANOID MAP

Ian Byrne, 'Soviet Tourist Maps: A Short Overview', *The Cartographic Journal*, Vol. 59, Issue 4, 2022, pp. 395–404

John Davies and Alexander J. Kent, *The Red Atlas: How the Soviet Union Secretly Mapped the World*, University of Chicago Press, 2017

Bill Keller, 'Soviet Aide Admits Maps Were Faked for 50 Years', *New York Times*, 3 September 1988, https://www.nytimes.com/1988/09/03/world/soviet-aide-admits-maps-were-faked-for-50-years.html (accessed December 2024)

Michael Parks, 'Moscow to Place Nation on Right Road After Faking Maps for 50 years', *Los Angeles Times*, 4 September 1988, https://www.latimes.com/archives/la-xpm-1988-09-04-mn-2289-story.html (accessed December 2024)

Alexey V. Postnikov, 'Maps for Ordinary Consumers versus Maps for the Military: Double Standards of Map Accuracy in Soviet Cartography, 1917–1991', *Cartography and Geographic Information Science*, Vol. 29, Issue 3, 2002, pp. 243–60

Dagmar Unverhau (ed.), *State Security and Mapping in the German Democratic Republic: Map Falsification as a Consequence of Excessive Secrecy?*, Lectures to the Conference of the BStU, Berlin, 8–9 March 2001, Transaction Publishers, 2006

10 THE DEADLY SHORTCUT

Larry Bishop, 'Researcher's Guide to Sutter's Fort's Collections of Donner Party Material', California State Parks, November 2005, https://www.parks.ca.gov/pages/485/files/Donner_Party_Material_Master_Guide_April_2008.pdf (accessed January 2025)

Lansford W. Hastings, *The Emigrants' Guide to Oregon and California* (1845), Applewood Books, 1994

Heinrich Lienhard, 'The Journal of Heinrich Lienhard, July 26–September 8, 1846', *Utah Historical Quarterly*, Vol. 19, Nos 1–4, 1951, https://issuu.com/utah10/docs/volume_19_1951/s/88944 (accessed January 2025)

Peter Meyerhoff, 'The Hastings Overland Party of 1845, and the Genesis of the Cutoff' (lecture), 2021, https://www.youtube.com/watch?v=NfHkmZkRr-o (accessed January 2025)

Hiram Miller and James Reed, diaries entries, 1846, https://www.donnerpartydiary.com/jul46.html (accessed January 2025)

Virginia Reed Murphy, 'Across the Plains in the Donner Party: A Personal Narrative of the Overland Trip to California', *Century Magazine*, Vol. 42, 1891

Ethan Rarick, *Desperate Passage: The Donner Party's Perilous Journey West*, Oxford University Press, 2008

11 THE MAP THAT'S BLANK (WHERE THE STREETS HAVE NO NAME)

Dharmesh Ba and Gowri N. Kishore, 'Anatomy of an Indian Address', *The India Notes*, 8 August 2023, https://newsletter.theindianotes.com/p/anatomy-of-an-indian-address (accessed June 2025)

Clint Eastwood et al., 'When did streets get names?', StackExchange, 30 June 2014, https://history.stackexchange.com/questions/14390/when-did-streets-get-names (accessed June 2025)

Elizabeth Laraki, 'Google Maps UX: The India Conundrum – How Google Maps first failed and then creatively adapted to flourish in India', Substack, 5 September 2024, https://elizlaraki.substack.com/p/google-maps-ux-the-india-conundrum (accessed June 2025)

Deirdre Mask, *The Address Book: What Street Addresses Reveal about Identity, Race, Wealth and Power*, Profile Books, 2020

Malavika Neurekar, 'Politics and emotions behind renaming Mumbai's streets', *Hindustan Times*, 19 November 2018, https://www.hindustantimes.com/mumbai-news/what-s-in-a-name-you-ask-politics-emotions-legacy/story-MpYpqkId5Yk9I6JEMj5c5O.html (accessed June 2025)

Mark Sappenfield, 'Tear up the maps: India's cities shed colonial names', *Christian Science Monitor*, 17 September 2006, https://www.csmonitor.com/2006/0907/p01s02-wosc.html

Balaji Viswanathan et al., 'Why are the maps of India in Google Maps or any other online maps not so clear compared to western countries?', Quora.com, n.d., https://www.quora.com/unanswered/Why-are-the-maps-of-India-in-Google-Maps-or-any-other-online-maps-not-so-clear-compared-to-western-countries (accessed June 2025)

'Welcome to India, where the streets have four names', *The Economist*, 5 September 2024, https://www.economist.com/asia/2024/09/05/welcome-to-india-where-the-streets-have-four-names (accessed June 2025)

Alejandro Zúñiga, 'Why doesn't Costa Rica use real addresses? And who cares about an old fig tree, anyway?', *Costa Rica Daily*, 18 February 2021, https://www.crcdaily.com/p/why-doesnt-costa-rica-use-real-addresses (accessed June 2025)

12 THE MAP THAT SHOULD HAVE KNOWN BETTER

'The Guardian view on knowledge in an information age: take it to heart' (editorial), *Guardian*, 21 January 2016, https://www.theguardian.com/commentisfree/2016/jan/21/the-guardian-view-on-knowledge-in-an-information-age-take-it-to-heart (accessed June 2025)

Ken Jennings, *Maphead: Charting the Wide, Weird World of Geography Wonks*, Scribner, 2011

Simon Maloy, 'Fox News Discovers Nuclear Reactor in Japanese Disco', *Media Matters for America*, 14 March 2011, https://www.mediamatters.org/fox-news/fox-news-discovers-nuclear-reactor-japanese-disco (accessed February 2025)

'National Geographic-Roper Public Affairs 2006 Geographic Literacy Study', *National Geographic*, May 2006, https://media.nationalgeographic.org/assets/file/NGS-Roper-2006-Report.pdf (accessed February 2025)

'News outlet mistakes New Zealand for Japan in map bungle', 9News, 26 August 2019, https://www.9news.com.au/world/world-news-new-zealand-mistaken-for-japan-in-map-mistake-jacinda-ardern/545407b0-c7d9-4385-adeb-77ec28bd3f11 (accessed January 2025)

Bijal P. Trivedi, 'Survey Reveals Geographic Illiteracy', *National Geographic*, 20 November 2002 https://www.nationalgeographic.com/science/article/geography-survey-illiteracy

U.S. Adults' Knowledge About the World, Gallup, Council on Foreign Relations and *National Geographic*, December 2019, https://www.cfr. org/report/us-adults-knowledge-about-world (accessed June 2025)

What College-Aged Students Know About the World: A Survey on Global Literacy, Council on Foreign Affairs and *National Geographic*, September 2016, https://cdn.cfr.org/sites/default/files/pdf/cfr_natgeo_ asurveyongloballiteracy.pdf (accessed June 2025)

'Where in the World? A Global Look at Geographic Recognition', Holiday Cottages, n.d., https://www.holidaycottages.co.uk/where-in-the-world-is/ (accessed January 2025)

Andrew Wiseman, 'When Maps Lie: Tips from a geographer on how to avoid being fooled', BloombergUK, 25 June 2015, https://www. bloomberg.com/news/articles/2015-06-25/how-to-avoid-being-fooled-by-bad-maps (accessed February 2025)

13 THE MAP THAT BROKE THE FRAME

Jerry Brotton, *A History of the World in Twelve Maps*, Penguin, 2013

Eric Carle, *The Very Hungry Caterpillar*, Philomel Books, 1969

Arthur Davies, 'Behaim, Martellus and Columbus', *Geographical Journal*, Vol. 143, No. 3, November 1977, pp. 451–9, https://www.jstor.org/ stable/634713 (accessed June 2025)

Árpád Papp-Váry, 'A pálcikatérképtöl az ürtérképig' ('From the Stick Chart to the Space Map'), in István Klinghammer and Árpád Papp-Váry (eds), *Földünk tükre a térkép* (*The Map as a Mirror of the Earth*), Gondolat Könyvkiadó, 1983, p. 15

William Richardson, 'South America on Maps Before Columbus? Martellus's "Dragon's Tail" Peninsula', *Imago Mundi*, Vol. 55, No. 1, 2003, pp. 25–37

Jim Seibold, 'Martellus' World Maps 1489–1490' and 'Behaim Globe', www.myoldmaps.com (accessed February 2025)

14 THE UNREADABLE MAP (AND THE MAP THAT BLEW ITSELF UP)

Bikini Atoll: Nomination by the Republic of the Marshall Islands for Inscription on the World Heritage List 2010, January 2009, https://whc.unesco.org/uploads/nominations/1339.pdf (accessed June 2025)

William Davenport, 'Maps and Mapmaking: Marshall Island Stick Charts', in Helaine Selin (ed.), *Encyclopaedia of the History of Science, Technology, and Medicine in Non-Western Cultures*, Springer Nature, 2008, pp. 1318–20

Kenneth O. Emery, J. I. Tracey, Jr. and H. S. Ladd, 'Geology of Bikini and Nearby Atolls – Bikini and Nearby Atolls: Part 1, Geology', *Geological Survey Professional Paper 260-A*, U.S. Department of the Interior, 1954

Joseph H. Genz, 'Navigating the Revival of Voyaging in the Marshall Islands: Predicaments of Preservation and Possibilities of Collaboration', *The Contemporary Pacific*, Vol. 23, No. 1, March 2011, pp. 1–34

Joseph H. Genz, *Breaking the Shell: Voyaging from Nuclear Refugees to People of the Sea in the Marshall Islands*, University of Hawai'i Press, 2018

Keith M. Parsons and Robert A. Zaballa, *Bombing the Marshall Islands: A Cold War Tragedy*, Cambridge University Press, 2017

Serhii Plokhy, *Atoms and Ashes: From Bikini Atoll to Fukishima*, Allen Lane, 2022

Sam Scott, 'What Bikini Atoll Looks Like Today', *Stanford Magazine*, December 2017, https://medium.com/stanford-magazine/stanford-research-on-effects-of-radioactivity-from-bikini-atoll-nuclear-tests-on-coral-and-crab-dna-48459144020c (accessed December 2024)

Leonard P. Schultz, 'Field charts of Marshall Islands: Kwajalein Atoll, eastern part, 1944, Bikini Atoll, annotations by Schultz, 1946, Rongelap Atoll, Rongerik Atoll', Smithsonian Institution Archive, 1944

Dirk H. R. Spennemann, 'Traditional Marshallese Stickchart Navigation', in Dirk H. R. Spennemann, *Essays on the Marshallese Past*, Albury,

1998, https://marshall.csu.edu.au/Marshalls/html/essays/es-tmc-2.html (accessed January 2025)

Kim Tingley, 'The Secrets of the Wave Pilots', *New York Times*, 17 March 2016, https://www.nytimes.com/2016/03/20/magazine/the-secrets-of-the-wave-pilots.html (accessed December 2024)

J. I. Tracey, H. S. Lapham and J. E. Hoffmeister, 'Reefs of Bikini, Marshall Islands', *Geological Society of America Bulletin*, Vol. 59, No. 9, 1948, pp. 861–98

Waan Aelōñ in Majel programme website, https://www.canoesmarshallislands.com/ (accessed January 2025)

15 THE WORLD MAP THAT WASN'T

Jerry Brotton, *Great Maps: The World's Masterpieces Explored and Explained*, Dorling Kindersley, 2014

Georgia Brown, *International Map of the World*, ArcGIS, https://storymaps.arcgis.com/stories/3246c301f549450c9a53caf89d293c72 (accessed March 2025)

Alistair Pearson and Michael Heffernan, 'Globalising Cartography? The International Map of the World, the International Geographical Union, and the United Nations', *Imago Mundi*, Vol. 67, Issue 1, January 2015, pp. 58–80

Albrecht Penck, 'Construction of a map of the world on a scale of 1:1 million', *Geographical Journal*, Vol. 1, 1893, pp. 253–61

Steven Seegel, *Map Men: Transnational Lives and Deaths of Geographers in the Making of East Central Europe*, University of Chicago Press, 2018

16 THE MAP THAT'S NEVER WRONG

'1986: COMPACT DISCS – The future of CAR NAVIGATION?', *Top Gear*, BBC Archive, https://www.youtube.com/watch?v=qBCPAlGTnK0 (accessed June 2025)

Chris Brown, 'EY study: Over a third of UK consumers keen on a New Year "digital detox"', ey.com. 10 January 2025, https://www.ey.com/en_uk/newsroom/2025/01/over-a-third-of-uk-consumers-keen-on-a-new-year-digital-detox (accessed May 2025)

Calum Cockburn, 'The maps of Matthew Paris', British Library Medieval manuscripts blog, 1 August 2020, https://blogs.bl.uk/digitisedmanuscripts/2020/07/the-maps-of-matthew-paris.html (accessed May 2025)

'In-Car Sat Nav? Without a Satellite? In 1971?' (video documentary), *Tomorrow's World*, BBC Archive, https://www.youtube.com/watch?v=4qqnHtH1RAs (accessed June 2025)

Eleanor Maguire et al., 'Navigation-related structural change in the hippocampi of taxi drivers', *Proceedings of the National Academy of Sciences of the United States of America*, Vol. 97, No. 8, 14 March 2000, pp. 4398–403

'The surprising social trend among young adults in 2025: McCrindle Research survey reveals shift in tech, work and shopping views', *Adelaide Now*, 2025

'Trends of 2025 report', McCrindle Research, https://mccrindle.com.au/resource/report/trends-of-2025-report/ (accessed May 2025)

OTHER SOURCES CONSULTED

Edward Brooke-Hitching, *The Phantom Atlas The Greatest Myths, Lies and Blunders on Maps*, Simon & Schuster, 2016

Mark's neighbour John, who knows loads of stuff (accessed March 2022 – October 2025)

Jessica McFadyen and Oliver Baumann, 'Analysis: It's not just doorways that make us forget what we came for in the next room', The Conversation, 9 March 2021, https://theconversation.com/its-not-just-doorways-that-make-us-forget-what-we-came-for-in-the-next-room-156030 (accessed June 2025)

Mark Monmonier, *How to Lie with Maps*, University of Chicago Press, 2018

Mark Monmonier, *How to Lie with Maps*, University of Chicago Press, 2018 (we read it twice)

Sumathi Reddy, 'Want to Lower Your Alzheimer's Risk? Taxi Drivers Offer a Clue', *Wall Street* Journal, 24 December 2024, https://www.wsj.com/health/wellness/alzheimers-risk-taxi-ambulance-drivers-be15739b?utm (accessed January 2025)

PICTURE CREDITS

pp. 279, 285 U.S. Geological Survey and Association of American State Geologists, 2024. The National Geologic Map Database: U.S. Geological Survey National Geologic Map Database website, https://ngmdb.usgs.gov/

p. 295 © Board of Trustees of the British Museum

p. 302 Copernicus Sentinel Data 2017/Orbital Horizon/Gallo Images/Getty Images

p. 304 (left) © Board of Trustees of the Science Museum

p. 304 (right) Shutterstock.com

p. 308 Norman B. Leventhal Map and Education Center, Boston Public Library

p. 319 The American Geographical Society Library, University of Wisconsin-Milwaukee Libraries

p. 331 British Library Archive/Bridgeman Images

PLATE SECTION

2. Library of Congress, Geography and Map Division, Louisiana: European Explorations and the Louisiana Purchase

3. Beinecke Rare Book and Manuscript Library, Yale University

4. John Cary, public domain via Wikimedia Commons

5. Thomas J. Maslen, public domain via Wikimedia Commons

6. National Library of Scotland

7. Courtesy Dr Alex Kent

8. Library of Congress, Geography and Map Division

9. David Rumsey Map Collection

10. The History Collection/Alamy Stock Photo

11. Jimlop Collection/Alamy Stock Photo

12. Library of Congress, Geography and Map Division

13. U.S. Geological Survey and Association of American State Geologists, 2024. The National Geologic Map Database: U.S. Geological Survey National Geologic Map Database website, https://ngmdb.usgs.gov/

14. Copernicus Sentinel Data 2017/Orbital Horizon/Gallo Images/Getty Images

15. The American Geographical Society Library, University of Wisconsin-Milwaukee Libraries

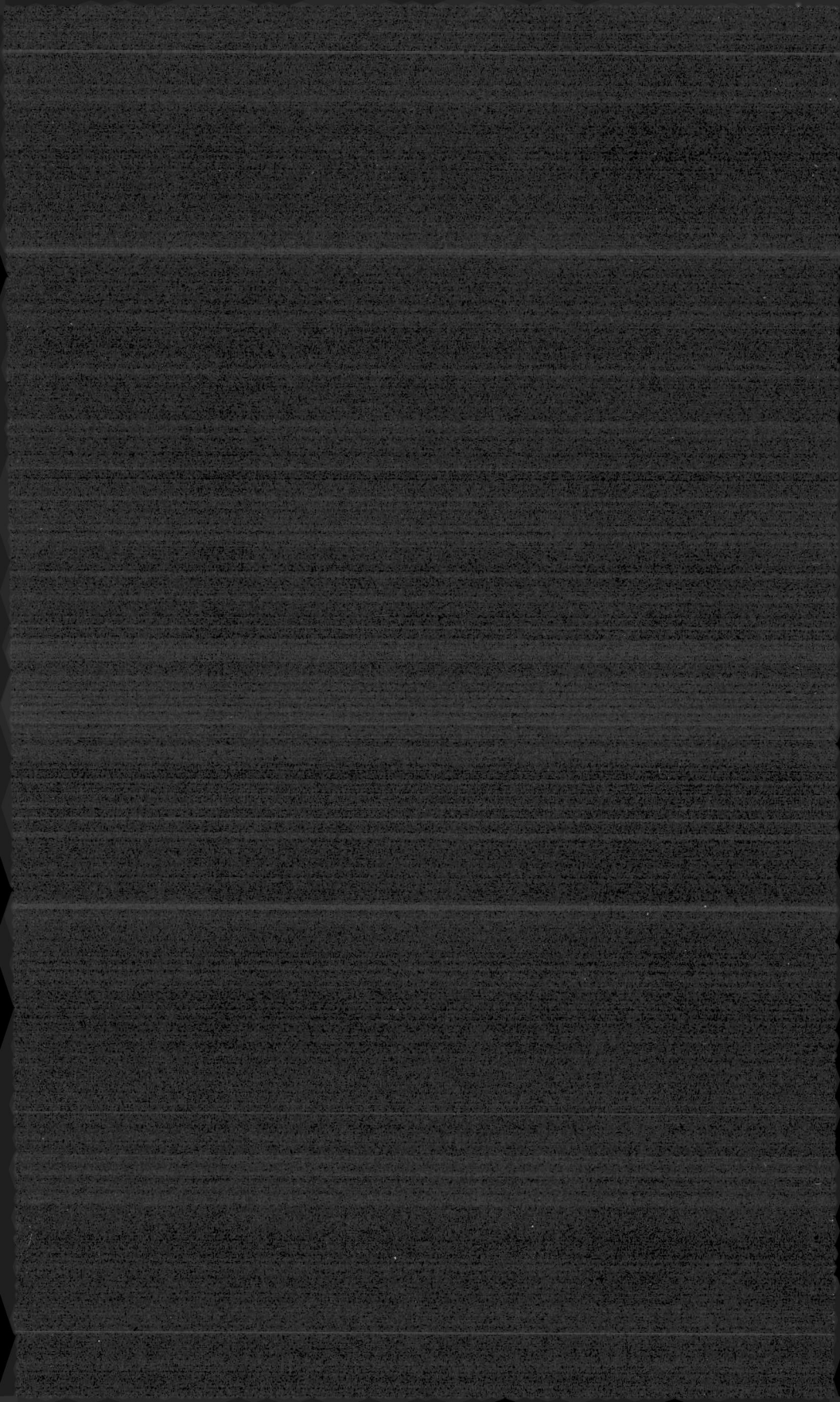